SpringerBriefs in Mathematics

SpringerBriefs in Mathematics showcases expositions in all areas of mathematics and applied mathematics. Manuscripts presenting new results or a single new result in a classical field, new field, or an emerging topic, applications, or bridges between new results and already published works, are encouraged. The series is intended for mathematicians and applied mathematicians.

More information about this series at http://www.springer.com/series/10030

Yoshihiro Tonegawa

Brakke's Mean Curvature Flow

An Introduction

Springer

Yoshihiro Tonegawa
Tokyo Institute of Technology
Tokyo, Japan

ISSN 2191-8198   ISSN 2191-8201   (electronic)
SpringerBriefs in Mathematics
ISBN 978-981-13-7074-8   ISBN 978-981-13-7075-5   (eBook)
https://doi.org/10.1007/978-981-13-7075-5

Library of Congress Control Number: 2019934770

This Springer imprint is published by the registered company Springer Nature Singapore Pte Ltd.
The registered company address is: 152 Beach Road, #21-01/04 Gateway East, Singapore 189721,
Singapore

To Lili, Sana, Suzue, and Yoshitsugu.

Preface

The aim of this short book is to introduce to graduate students and researchers a notion of mean curvature flow created by Ken Brakke in [7], which is the expanded book version published in 1978 of his Ph.D. thesis under his supervisor Fred Almgren. At present, this notion of mean curvature flow, which is often called the *Brakke flow*, is little known outside of a small circle of specialists in geometric measure theory, and my hope is that this book gives a relatively easy and direct pathway to the Brakke flow without going through a number of heavy-handed proofs of preliminary materials. For that purpose, the first part covers a minimum amount of prerequisite from geometric measure theory, often with intuitive descriptions and examples. I have given a series of lectures on this subject aimed at graduate students and young post-docs at Fourier Institute, UC Berkeley, Kyoto University, ICTP Trieste, and Hausdorff Research Institute in recent years, and the lecture notes for these occasions served as a good preparation for the present book.

It was my fortune that I encountered the book [7] in the early days of my career and I would like to describe how that happened. I heard about the book for the first time in Craig Evans' lecture given at Rice University in 1995, where I was a postdoctoral instructor at the time. He gave a series of lectures on the level set method for the mean curvature flow, in which he explained his joint work with Spruck [14] connecting the level set solution to the Brakke flow. One thing which was surprising and even mysterious to me was the regularity theory in [7]: Craig mentioned that almost all level sets of his solution established in [13] should be almost everywhere and for almost all time smooth according to Brakke's regularity theory. I had learned about the corresponding regularity theory for minimal surfaces, but Brakke's formulation for mean curvature flow differs greatly from it, and I wanted to understand the mechanism of this regularity theory. Soon after, I spent many days reading [7] in detail trying to understand the proof. I also ran a weekly reading seminar for a while with Bob Hardt, who was my excellent mentor at Rice. After some months, however, I felt that I could not follow the proof and the long chain of computations. The book contained a wealth of great ideas, but the proof was simply impenetrable for me. I went on and worked on different problems, in particular on problems involving "diffused interfaces," which are indirectly related to the Brakke flow. As an application of this approach, in [27], we proved some existence theorems for "Brakke-like" flows, where the velocity of the moving interface is equal to the mean curvature plus given irregular vector fields. When I started thinking about the regularity theorem for this flow with my student, we soon realized that the line of proof given by Brakke in [7] would not work for more general flows such as the one in [27]. This realization eventually led us to give a different approach to the proof of regularity for the Brakke flow. So after more than 15 years, serendipity led us to a rather good understanding of Brakke's regularity theory described in [21, 37] and a few years later of his existence theory described in [22]. Beside being an introduction to the Brakke flow, explaining our results in [21, 22, 37] with some intuitive outline of proofs is the goal of this book.

The notion of mean curvature flow (abbreviated hereafter as MCF) itself has no ambiguity if it is a classical solution, i.e., given a smooth compact k-dimensional surface $\Gamma_0 \subset \mathbb{R}^n$ ($1 \leq k < n$), there exists a unique family $\{\Gamma(t)\}_{t \in [0,T)}$ of smooth compact k-dimensional surfaces for some $T < \infty$ with $\Gamma(0) = \Gamma_0$ and the normal velocity of $\Gamma(t)$ equal to the mean curvature of $\Gamma(t)$ at each point $x \in \Gamma(t)$ and time $t \in [0, T)$. For example, if $k = n - 1$, and if $\Gamma(t)$ is represented locally as a graph $\{(x, f(x, t)) \in \mathbb{R}^n : x \in U \subset \mathbb{R}^{n-1}\}$, one can check that the function f should satisfy

$$
\frac{f_t}{\sqrt{1 + |\nabla f|^2}} = \mathrm{div} \left(\frac{\nabla f}{\sqrt{1 + |\nabla f|^2}} \right)
$$

$$
= \frac{1}{\sqrt{1 + |\nabla f|^2}} \left(\Delta f - \sum_{i,j=1}^{n-1} \frac{f_{x_i} f_{x_j}}{1 + |\nabla f|^2} f_{x_i x_j} \right)
$$

where $f_t = \frac{\partial f}{\partial t}$, $\nabla f = (f_{x_1}, \ldots, f_{x_{n-1}})$, and $f_{x_i} = \frac{\partial f}{\partial x_i}$. The left-hand side is the normal velocity and the right-hand side is the mean curvature of $\Gamma(t)$. Given any point of the surface, we may choose an appropriate coordinate system so that $|\nabla f|^2$ is seen to be a small quantity in a small space–time neighborhood. With such a coordinate system, we have $f_t \approx \Delta f$, so that the equation is expected to share the same property of the heat equation. The MCF has a distinctive feature that the k-dimensional surface area of $\Gamma(t)$ decreases as t increases. Because of the nonlinearity, the family of surfaces moving by MCF may develop some singularity in finite time denoted here by T. What Brakke envisioned as a MCF, however, is a much broader class of flows which can move *with singularities* such as networks of curves joined by junctions on the plane. Physically, such flows with singularities are observed in motion of grain boundaries in a pure metal during the annealing process. The presence of boundaries increases the total energy of the system due to the surface tension, and the boundaries move to reduce the area of boundaries, resulting in the MCF. To treat such phenomena, we need to broaden the class of k-dimensional surfaces from smooth categories to the one with singularities. A theory of surfaces inclusive of such singularities was earlier introduced by Almgren [3], which Almgren called *varifolds* (a name created from "variational manifold"), and

Brakke's notion of MCF is considered in the space of varifolds. It strikes me as amazing that such a general framework was considered right from the beginning of the study of MCF. Note that little was known about even smooth MCF at the time (back in the 1970s, preceding works of Ecker, Huisken, and numerous others), and I am sure that anyone would start out with a smooth case instead of diving into the world of varifolds, struggling squarely with singularities! So, even though I feel that I have a better understanding of most of what Brakke wanted to accomplish in [7], I do not understand how one could even think of what is written in it at such an early stage and I have a sense of awe for the accomplishment. I hope that this feeling can be partly shared by the reader of this book.

To render the book a short and comprehensible introduction, I exclusively discuss the Brakke flow through the narrow scope of my own research area, and not numerous other types or aspects of MCF. For additional information on the Brakke flow, see [19, 43]. There are numerous books and survey articles on the MCF available, see for example [5, 10, 11, 15, 28, 29] and the references therein.

The organization of this book is as follows. In Chap. 1, a number of tools from geometric measure theory are presented with intuitive explanations and mostly without proofs. Readers unfamiliar with geometric measure theory may gain some intuition and ideas to understand the setting of Brakke flow. In Chap. 2, the notion of generalized velocity for varifolds is discussed and the Brakke flow is defined. In Chap. 3, some preliminary properties of the Brakke flow are presented. They are about certain compactness theorems and existence of tangent flow. The result in Chap. 4 establishes a general existence theorem for the Brakke flow starting from singular hypersurfaces. The content of this chapter is a thorough reworking of [7, Chapter 4] and it is intended as an introduction to our work [22]. The content of Chap. 5 is an introduction to the Allard regularity theorem [1]. This is intended as a prelude to a more complicated time-dependent counterpart of the subsequent chapter. While some parts of the proof such as construction of the Lipschitz approximation are omitted, the essential part of the regularity proof is given so the reader has some feeling of what the ingredients are. Chapter 6 explains the regularity theorem for the Brakke flow, intended as an introduction to [21, 37]. These papers are again thorough reworkings of [7, Chapter 6] with a new approach. Just like his existence theorem, for many years Brakke's original proof had remained poorly understood and the complete proof using different techniques was first given in these papers.

I would like to express my gratitude to Mr. Masayuki Nakamura of Springer Japan for asking me to write and for patiently waiting for the manuscript. I acknowledge the support of JSPS grant-in-aid for scientific research (A) #25247008 and (S) #21224001 for carrying out my research activities.

Tokyo, Japan   Yoshihiro Tonegawa

Contents

Chapter 1
Preliminary Materials

1.1 Basic Notation

Throughout, $1 \leq k < n$ are integers, $\mathbb{N}$ is the set of natural numbers, $\mathbb{R}^n$ is the n-dimensional Euclidean space and

$$\mathbb{R}^+ := \{x \in \mathbb{R} : x \geq 0\}.$$

Let $U \subset \mathbb{R}^n$ be an open set. The symbol

$$C_c^l(U)$$

denotes the set of l times continuously differentiable functions with compact support in U and

$$C_c(U) := C_c^0(U).$$

A function $f : U \to \mathbb{R}$ is said to be Lipschitz if

$$\mathrm{Lip}(f) := \sup_{x,y \in U} \frac{|f(x) - f(y)|}{|x - y|} < \infty.$$

The set of vector fields with each component in $C_c^l(U)$ is denoted by

$$C_c^l(U; \mathbb{R}^n)$$

and

$$C_c(U; \mathbb{R}^n) := C_c^0(U; \mathbb{R}^n).$$

© The Author(s), under exclusive license to Springer Nature Singapore Pte Ltd. 2019
Y. Tonegawa, *Brakke's Mean Curvature Flow*, SpringerBriefs in Mathematics,
https://doi.org/10.1007/978-981-13-7075-5_1

For $g \in C_c^1(U; \mathbb{R}^n)$, we define ∇g as the $n \times n$ matrix-valued function whose first row is composed of partial derivatives of the first component of g and so forth. By $\circ$, we indicate the usual matrix multiplication.

Set

$$B_r^n(x) := \{y \in \mathbb{R}^n : |y - x| \leq r\}$$

and we often write $B_r(x)$ when the dimension is clear from the context and B_r for $B_r(0)$. We write

$$\mathcal{H}^k$$

for the k-dimensional Hausdorff measure on $\mathbb{R}^n$. For the reader unfamiliar with the Hausdorff measure, it is more or less enough to think of $\mathcal{H}^k$ as the "measure of k-dimensional surface area", i.e., whenever a piece of k-dimensional surface Γ is given, $\mathcal{H}^k(\Gamma)$ is the k-dimensional surface area. The class of $\mathcal{H}^k$ measurable subsets includes all Borel sets. The measure $\mathcal{H}^k$ is equipped with the usual nice properties such as the countable additivity for the measurable sets just like the Lebesgue measure on $\mathbb{R}^n$, which is denoted by

$$\mathcal{L}^n.$$

On $\mathbb{R}^n$, $\mathcal{H}^n$ and $\mathcal{L}^n$ coincide as measures. To gain more information on and familiarity with the Hausdorff measure, see [12, 33]. We define

$$\omega_k := \mathcal{L}^k(B_1^k).$$

By the term *Radon measure on U*, we mean a Borel regular measure which is also locally finite. Given a Radon measure μ on U, the support of μ is defined by

$$\operatorname{spt} \mu := \{x \in U : \mu(B_r(x)) > 0 \text{ for all } B_r(x) \subset U\}.$$

The restriction of measure μ to a subset Γ is denoted by

$$\mu \llcorner_\Gamma .$$

For short, we often write

$$\mu(\phi) := \int_U \phi(x) \, \mathrm{d}\mu(x).$$

For $1 \leq p < \infty$,

$$L^p(\mu)$$

denotes the p-th power integrable class of μ-measurable function $\overset{\cdot}{f}$, i.e., $\int_U |f|^p \, d\mu < \infty$.

$$L^p_{loc}(\mu)$$

denotes the locally p-th power integrable class.

We write

$$\mathbf{G}(n, k)$$

for the set of all k-dimensional subspaces in $\mathbb{R}^n$. For example, $\mathbf{G}(2, 1)$ is the set of lines passing through the origin. Later we also use the set of all k-dimensional affine planes in $\mathbb{R}^n$ denoted by

$$\mathbf{A}(n, k).$$

For each $S \in \mathbf{G}(2, 1)$, if we let $\theta \in [0, \pi)$ be the counterclockwise angle between the positive x-axis and S, we may identify $\mathbf{G}(2, 1)$ with $[0, \pi)$ through this correspondence. A k-dimensional subspace $S \in \mathbf{G}(n, k)$ is often identified with the $n \times n$ matrix representing the orthogonal projection $\mathbb{R}^n \to S$. Such a projection matrix can be obtained as follows. Let $v_1, \ldots, v_k \in \mathbb{R}^n$ be an orthonormal basis of S. Consider the $n \times n$ matrix

$$\sum_{j=1}^{k} v_j \otimes v_j.$$

Here, $v \otimes v$ is the $n \times n$ matrix with its (l, m) component given by the product of l-th and m-th components of v. One can see that this matrix is symmetric and is the identity map on S, and the kernel is the orthogonal complement of S, which is denoted by

$$S^\perp \in \mathbf{G}(n, k - n).$$

For example, we may identify

$$\mathbf{G}(2, 1) = \left\{ \begin{pmatrix} \cos^2 \theta & \sin \theta \cos \theta \\ \sin \theta \cos \theta & \sin^2 \theta \end{pmatrix} : \theta \in [0, \pi) \right\}.$$

For two elements A and B of $\mathrm{Hom}(\mathbb{R}^n; \mathbb{R}^n)$ which is the space of all $n \times n$ matrices, define a scalar product

$$A \cdot B := \mathrm{tr}(A^\top \circ B)$$

where $A^\top$ is the transpose of A. The identity of $\mathrm{Hom}(\mathbb{R}^n; \mathbb{R}^n)$ is denoted by I. For $A \in \mathrm{Hom}(\mathbb{R}^n; \mathbb{R}^n)$ define

$$|A| := \sqrt{A \cdot A}, \quad \|A\| := \sup\{|A(x)| : x \in \mathbb{R}^n, |x| = 1\}.$$

They define the same topology on $\mathrm{Hom}(\mathbb{R}^n; \mathbb{R}^n)$. As a subspace, $\mathbf{G}(n, k) \subset \mathrm{Hom}(\mathbb{R}^n; \mathbb{R}^n)$ is compact with this topology. We often think of the codimension one case, $\mathbf{G}(n, n-1)$. For any $S \in \mathbf{G}(n, n-1)$, there is a unit vector v perpendicular to S. The matrix representing the orthogonal projection to S is then

$$I - v \otimes v.$$

Through this correspondence $S \leftrightarrow \pm v$, $\mathbf{G}(n, n-1)$ can be homeomorphically identified with $\mathbb{S}^{n-1} / \pm 1$, the $n-1$-dimensional real projective space.

1.2   Countably Rectifiable Sets

By saying that $\Gamma \subset \mathbb{R}^n$ is a k-dimensional C^1 surface, we mean that for any $x \in \Gamma$, there exist an open set $U \subset \mathbb{R}^n$ containing x, an open set $\tilde{U} \subset \mathbb{R}^k$ and a proper injective C^1 function $f : \tilde{U} \to U$ such that

$$f(\tilde{U}) = \Gamma \cap U \text{ with } |\nabla f| \neq 0 \text{ on } \tilde{U}.$$

The classical k-dimensional tangent space denoted by $\mathrm{T}_y \Gamma$ is defined as the image of $\nabla f(x) \in \mathrm{Hom}(\mathbb{R}^k; \mathbb{R}^n)$ and with $y = f(x)$. We naturally think of $\mathrm{T}_y \Gamma$ as an element of $\mathbf{G}(n, k)$. More generally, we would like to treat subsets which are like C^1 surfaces measure-wise and which are not necessarily C^1 everywhere. For example, we would like to treat "networks" or "soap bubble clusters" which have singularities typically with lower dimensions, or more precisely, singularities of null k-dimensional measure. For this purpose, we define the following:

Definition 1.1 A subset $\Gamma \subset \mathbb{R}^n$ is called **countably k-rectifiable** if there exists a sequence of k-dimensional C^1 surfaces $\Gamma_1, \Gamma_2, \cdots$ such that $\mathcal{H}^k(\Gamma \setminus \cup_{j=1}^\infty \Gamma_j) = 0$.

Any k-dimensional C^1 surface is countably k-rectifiable, since one can take the surface itself as Γ_1. So is any countable union of k-dimensional C^1 surfaces, since they can be taken as Γ_j. These C^1 surfaces Γ_j need not be disjoint. They can intersect each other and thus there may be singularities such as corners and cusps. The definition requires only *inclusion*, thus *any* subset of a countably k-rectifiable set is again countably k-rectifiable. Note also that a countable union of countably k-rectifiable sets is also countably k-rectifiable by diagonal argument. The last property is generally desirable for various measure-theoretic operations. We mention that there is another equivalent definition of being countably k-rectifiable using Lipschitz functions, expressed as follows.

Proposition 1.2 *A subset $\Gamma \subset \mathbb{R}^n$ is countably k-rectifiable if and only if there exists a sequence of Lipschitz functions $f_j : \mathbb{R}^k \to \mathbb{R}^n$ such that $\mathcal{H}^k(\Gamma \setminus \cup_{j=1}^\infty f_j(\mathbb{R}^k)) = 0$.*

Thus, instead of C^1, we may relax the regularity of the covering to Lipschitz. It is often convenient to take this "Lipschitz" version, since a Lipschitz function is more flexible than a C^1 function. For example, if $f : \mathbb{R}^n \to \mathbb{R}^n$ is a Lipschitz map and if $\Gamma \subset \mathbb{R}^n$ is countably k-rectifiable, $f(\Gamma)$ is also countably k-rectifiable.[1] A main reason for Proposition 1.2 to be true is the Rademacher theorem which gives almost everywhere differentiability of Lipschitz functions. The interested reader should consult [33] for the proof. A word of caution is that we may have $\overline{\Gamma}$, the topological closure of Γ, wildly different from Γ. Consider the following example. Let $\{x_j\}_{j \in \mathbb{N}}$ be a dense set of points on $\mathbb{R}^2$. Let Γ be the union of circles of radius $1/2^j$ centered at x_j. Then, Γ is countably 1-rectifiable and we have

$$\mathcal{H}^1(\Gamma) \leq \sum_{j \in \mathbb{N}} 2\pi/2^j = 2\pi.$$

On the other hand, since $\{x_j\}_{j \in \mathbb{N}}$ is dense, we can prove $\overline{\Gamma} = \mathbb{R}^2$. Thus, even if $\mathcal{H}^1(\Gamma) < \infty$, the topological closure $\overline{\Gamma}$ may not look like Γ at all in terms of measure or dimension. Equivalently, if we consider the Radon measure $\mu = \mathcal{H}^1 \llcorner_\Gamma$ with this Γ, by the same reason, we have spt $\mu = \mathbb{R}^2$, thus $\mathcal{H}^1(\text{spt}\,\mu \setminus \Gamma) = \infty$. So, when one takes the topological closure of a countably rectifiable set, one needs to be careful.

We next introduce a notion of measure-theoretic tangent space. Before we do so, let us define the following notation.

Definition 1.3 Given a set $\Gamma \subset U, x \in U$ and $\lambda > 0$, define

$$\Gamma_{x,\lambda} := \{z \in \mathbb{R}^n : z = \lambda^{-1}(y - x), y \in \Gamma\}. \tag{1.1}$$

As one can see from the definition, $\Gamma_{x,\lambda}$ is obtained by first shifting x to the origin, and then magnifying the set λ^{-1} times. If Γ is a k-dimensional C^1 surface and $x \in \Gamma$, then one can have a mental picture that $\Gamma_{x,\lambda}$ approaches to the tangent space $T_x\Gamma \in \mathbf{G}(n, k)$ locally uniformly as $\lambda \to 0+$. In our approach, we want to consider the convergence in measure, and we define the following measure-theoretic tangent space.

---

[1] This follows from a nice property of Lipschitz functions: for any set $A \subset \mathbb{R}^n$, we have $\mathcal{H}^k(f(A)) \leq (\text{Lip}(f))^k \mathcal{H}^k(A)$ (see [12]). Then it is easy to show the claim.

Definition 1.4 Suppose that $\Gamma \subset U$ is an $\mathcal{H}^k$ measurable set with locally finite $\mathcal{H}^k$ measure. We say that Γ has an **approximate tangent space** $S \in \mathbf{G}(n, k)$ at $x \in U$ if

$$\lim_{\lambda \to 0+} \mathcal{H}^k \llcorner_{\Gamma_{x,\lambda}} = \mathcal{H}^k \llcorner_S$$

as measures, that is, for all $\phi \in C_c(\mathbb{R}^n)$,

$$\lim_{\lambda \to 0+} \int_{\Gamma_{x,\lambda}} \phi(z) \, \mathrm{d}\mathcal{H}^k(z) = \int_S \phi(z) \, \mathrm{d}\mathcal{H}^k(z). \tag{1.2}$$

Here, the point is, instead of thinking of blown–up surfaces geometrically approaching to some subspace, we "test" the convergence of the blown–up surfaces to some subspace by integrating arbitrary but fixed continuous functions. Note that all information we need to know about a Radon measure is attainable as long as we know how it integrates continuous functions with compact support. If Γ is a k-dimensional C^1 surface and $x \in \Gamma$, then one can imagine (and prove) that the usual tangent space $T_x \Gamma$ is the approximate tangent space. With this definition, the $\mathcal{H}^k$ null set of Γ does not matter since it is defined through integration with respect to $\mathcal{H}^k$. When does such a tangent space not exist? If Γ is not C^1, and x happens to be on some "corner singularity" of Γ, then no such approximate tangent space can exist. If such an approximate tangent space exists at x, then it is unique. This is because, if it is not unique, then we would have two distinct $S, \tilde{S} \in \mathbf{G}(n, k)$ both satisfying (1.2) for any $\phi \in C_c(\mathbb{R}^n)$. Since they are different, we may certainly choose one particular $\phi \in C_c(\mathbb{R}^n)$ such that $\int_S \phi \, \mathrm{d}\mathcal{H}^k \neq \int_{\tilde{S}} \phi \, \mathrm{d}\mathcal{H}^k$. Then this is a contradiction to (1.2) since the left-hand side has to be unique by definition.

In general, given an $\mathcal{H}^k$ measurable set Γ with locally finite $\mathcal{H}^k$ measure, we cannot expect that *every point* of Γ has an approximate tangent space since there can be singularities such as corners. One may speculate on the other hand that the approximate tangent space may exist $\mathcal{H}^k$ a.e. on Γ, if not everywhere. Intuitively, singularities such as "corners" are lower dimensional, and such singular sets should have null $\mathcal{H}^k$ measure. This intuitively expected property turns out to be false in general. There is a class of $\mathcal{H}^k$ measurable sets with the property that approximate tangent space does not exist $\mathcal{H}^k$ a.e. on the set. In fact, we have the following beautiful result.

Proposition 1.5 *Suppose that $\Gamma \subset U$ is $\mathcal{H}^k$ measurable with locally finite $\mathcal{H}^k$ measure. Then, Γ is countably k-rectifiable if and only if Γ has the approximate tangent space for $\mathcal{H}^k$ a.e. $x \in \Gamma$.*

Thus, an $\mathcal{H}^k$ measurable set being countably k-rectifiable is completely characterized by having the approximate tangent space almost everywhere. In the following, we use the only if part, that is, any countably k-rectifiable set Γ has approximate tangent space $\mathcal{H}^k$ a.e. on Γ. The rough idea for the proof of the only if part is that Γ is included in a countable union of C^1 surfaces, and thus, in some suitable measure sense, any generic point of Γ is a "full k-dimensional density point" of some single C^1 surface disjoint from the rest of the C^1 surfaces. At such a point, the blow–up of Γ converges to the tangent space of that C^1 surface in measure. There are some measure-theoretic arguments to carry through this, but the details are omitted. The converse if part is harder to prove, since we need to make sure that Γ is included in a countable union of Lipschitz graphs (due to Proposition 1.2) almost everywhere, only from the existence of the approximate tangent space. The interested reader should see [33].

For an $\mathcal{H}^k$ measurable, locally finite countably k-rectifiable set Γ, we simply write $T_x\Gamma$ for the unique approximate tangent space at x with no fear of confusion. We also note that the correspondence $T_x\Gamma : \Gamma \to \mathbf{G}(n, k)$ is an $\mathcal{H}^k$ measurable function. Thus, for example, if we are given a function $\phi \in C(\mathbf{G}(n, k))$, $\int_\Gamma \phi(T_x\Gamma) \, d\mathcal{H}^k(x)$ is well-defined.

1.3 Varifolds

We next see the notion of a varifold. A varifold is a convenient setup which allows us to define the notion of mean curvature vector in a weak sense. The topology provided in the framework of a varifold is also of essential importance for us.

For an open set $U \subset \mathbb{R}^n$, we need first to consider the product space defined by

$$\mathbf{G}_k(U) := U \times \mathbf{G}(n, k)$$

with the usual product topology. Then, perhaps it is too terse to put this way, but let us start by saying that a **general k-varifold** V on U is a Radon measure on $\mathbf{G}_k(U)$. We denote the set of all k-varifolds on U by

$$\mathbf{V}_k(U).$$

We typically do not consider this completely general and abstract k-varifold, though it is important when we consider a convergence in this topology. The most important example of a k-varifold is the one naturally induced from a countably k-rectifiable set. More precisely, let $\Gamma \subset U$ be an $\mathcal{H}^k$ measurable, countably k-rectifiable set with locally finite $\mathcal{H}^k$ measure. As noted before, for $\mathcal{H}^k$ almost every $x \in \Gamma$, there exists a unique approximate tangent space $T_x\Gamma$. We want to define a Radon measure on $\mathbf{G}_k(U)$ from Γ. Due to the Riesz representation theorem, a Radon measure is completely determined once we give a locally bounded linear map from $C_c(\mathbf{G}_k(U))$ to $\mathbb{R}$. Given $\phi \in C_c(\mathbf{G}_k(U))$ whose independent variables are $x \in U$ and $S \in \mathbf{G}(n, k)$, the most natural thing to compute is

$$\int_\Gamma \phi(x, T_x\Gamma) \, d\mathcal{H}^k(x)$$

which is well-defined due to the existence of $T_x\Gamma$ on Γ almost everywhere. This assignment for $\phi \in C_c(\mathbf{G}_k(U))$ is linear, and for each compact set $K \subset U$ and for all ϕ with $\operatorname{spt}\phi \subset K \times \mathbf{G}(n,k)$,

$$\int_\Gamma \phi(x, T_x\Gamma)\,\mathrm{d}\mathcal{H}^k(x) \le \mathcal{H}^k(K \cap \Gamma) \sup_{x\in K,\, S\in\mathbf{G}(n,k)} |\phi(x,S)|,$$

so it defines a locally bounded linear functional. Thus we define

$$|\Gamma| \in \mathbf{V}_k(U)$$

by the rule

$$\int_{\mathbf{G}_k(U)} \phi(x,S)\,\mathrm{d}|\Gamma|(x,S) := \int_\Gamma \phi(x, T_x\Gamma)\,\mathrm{d}\mathcal{H}^k(x) \tag{1.3}$$

for all $\phi \in C_c(\mathbf{G}_k(U))$. Note that this is different from $\mathcal{H}^k\llcorner_\Gamma$, which is a Radon measure on U. Why do we want to consider such an abstract setting? There are many reasons. Suppose that we have a sequence of $\mathcal{H}^k$ measurable countably k-rectifiable sets Γ_j which have locally uniformly finite measures, i.e., for any compact set $K \subset U$, we have $\sup_j \mathcal{H}^k(\Gamma_j \cap K) < \infty$. This means that we have a sequence of Radon measures $\mathcal{H}^k\llcorner_{\Gamma_j}$ with locally uniformly finite measures. Then, by a general compactness theorem of Radon measures (see [12]), there exists a subsequence $\{\mathcal{H}^k\llcorner_{\Gamma_{j_l}}\}_{l\in\mathbb{N}}$ and a limit Radon measure μ on U such that

$$\lim_{l\to\infty} \int_U \phi(x)\,\mathrm{d}\mathcal{H}^k\llcorner_{\Gamma_{j_l}}(x) = \int_U \phi(x)\,\mathrm{d}\mu(x)$$

for all $\phi \in C_c(U)$. On the other hand, if we consider a sequence of varifolds $\{|\Gamma_j|\}_{j\in\mathbb{N}}$ instead, the sequence also has a locally uniformly finite measure on $\mathbf{G}_k(U)$. By the compactness theorem of Radon measures again, there exists a converging subsequence and a limit varifold $V \in \mathbf{V}_k(U)$ such that

$$\lim_{l\to\infty} \int_{\mathbf{G}_k(U)} \phi(x,S)\,\mathrm{d}|\Gamma_{j_l}|(x,S) = \int_{\mathbf{G}_k(U)} \phi(x,S)\,\mathrm{d}V(x,S)$$

for all $\phi \in C_c(\mathbf{G}_k(U))$. Since $\phi \in C_c(U)$ may be considered as an element of $C_c(\mathbf{G}_k(U))$ by letting $\phi(x,S) := \phi(x)$, we also have

$$\int_U \phi(x)\,\mathrm{d}\mu(x) = \int_{\mathbf{G}_k(U)} \phi(x)\,\mathrm{d}V(x,S).$$

From this, loosely speaking, we see that V contains more information than μ concerning in particular the behavior of tangent spaces of Γ_j. We will see more useful properties shortly.

For a general k-varifold $V \in \mathbf{V}_k(U)$, the **weight measure** $\|V\|$ which is a Radon measure on U is defined for all $\phi \in C_c(U)$ by

$$\int_U \phi(x) \, \mathrm{d}\|V\|(x) := \int_{\mathbf{G}_k(U)} \phi(x) \, \mathrm{d}V(x, S).$$

For $|\Gamma| \in \mathbf{V}_k(U)$, the weight measure $\||\Gamma|\|$ on U is

$$\int_U \phi(x) \, \mathrm{d}\||\Gamma|\|(x) = \int_{\mathbf{G}_k(U)} \phi(x) \, \mathrm{d}|\Gamma|(x, S) = \int_\Gamma \phi(x) \, \mathrm{d}\mathcal{H}^k(x),$$

thus we see that $\||\Gamma|\| = \mathcal{H}^k \llcorner \Gamma$. Though the class of k-varifolds of the form $|\Gamma|$ is fundamental, it is convenient to consider the following larger class of k-varifolds.

Definition 1.6 A k-varifold $V \in \mathbf{V}_k(U)$ is called **rectifiable** if there exist an $\mathcal{H}^k$ measurable countably k-rectifiable subset $\Gamma \subset U$ and an $\mathcal{H}^k$ measurable, locally $\mathcal{H}^k$ integrable function $\theta : \Gamma \to \mathbb{R}^+$ such that

$$\int_{\mathbf{G}_k(U)} \phi(x, S) \, \mathrm{d}V(x, S) = \int_\Gamma \phi(x, \mathrm{T}_x \Gamma) \theta(x) \, \mathrm{d}\mathcal{H}^k(x)$$

for all $\phi \in C_c(\mathbf{G}_k(U))$. The set of all rectifiable k-varifolds is denoted by

$$\mathbf{RV}_k(U).$$

For short, such a varifold is denoted by

$$\theta |\Gamma|.$$

If in addition

$$\theta(x) \in \mathbb{N}$$

for $\mathcal{H}^k$ a.e. $x \in \Gamma$, then V is called **integral**. The set of all integral k-varifolds is denoted by

$$\mathbf{IV}_k(U).$$

An example of a k-varifold ($k < n$) which is not rectifiable can be the following. Fix an arbitrary $S_0 \in \mathbf{G}(n, k)$ and define

$$V(\phi) := \int_U \phi(x, S_0) \, \mathrm{d}\mathcal{L}^n(x).$$

for $\phi \in C_c(\mathbf{G}_k(U))$. Note that U is not countably k-rectifiable, so V is not rectifiable. Intuitively, being rectifiable means that a measure is concentrated on some k-dimensional countably rectifiable set with the corresponding approximate tangent space. The difference between $|\Gamma|$ and this definition $\theta|\Gamma|$ is the presence of θ which takes non-negative values $\mathcal{H}^k$ a.e. on Γ. Of course, if $\theta = 1$, it is just $|\Gamma|$. The function θ is called **multiplicity**. The interpretation of θ for an integral varifold is that, if there is a portion of Γ on which $\theta = 2$, then one thinks that there are "two sheets" of k-dimensional sets there, and likewise for other integers. Allowing such multiplicity is convenient, since there may often be a situation that a sequence of two k-dimensional surfaces may come together in the limit, and we want to count the limit as a "two-sheeted" surface. For later use, we also define the "single-sheeted" surface as follows.

Definition 1.7 A k-varifold is said to be **unit density** if it is integral and the multiplicity is 1, i.e., $V = |\Gamma|$ for some $\mathcal{H}^k$ measurable countably k-rectifiable set Γ.

The class of unit density k-varifolds is the most natural measure-theoretic generalization of that of k-dimensional C^1 surfaces.

1.4 Mapping Varifolds

We next consider a mapping of a varifold. For a k-dimensional C^1 surface $\Gamma \subset U \subset \mathbb{R}^n$ and a smooth diffeomorphism $f : U \to U$, the image $f(\Gamma)$ is a C^1 surface and we should have an obvious correspondence of mapping of a varifold, $|\Gamma|$ to $|f(\Gamma)|$, which we may write

$$f_\sharp|\Gamma| := |f(\Gamma)|.$$

In terms of integration, by the change of variable formula, we have for $\phi \in C_c(\mathbf{G}_k(U))$

$$f_\sharp|\Gamma|(\phi) = |f(\Gamma)|(\phi) = \int_{f(\Gamma)} \phi(z, \mathrm{T}_z f(\Gamma))\, \mathrm{d}\mathcal{H}^k(z) \quad (z = f(x))$$

$$(1.4)$$

$$= \int_\Gamma \phi(f(x), \nabla f(x) \circ \mathrm{T}_x \Gamma)|\Lambda_k \nabla f(x) \circ \mathrm{T}_x \Gamma|\, \mathrm{d}\mathcal{H}^k(x).$$

Here $\nabla f(x) \circ \mathrm{T}_x \Gamma$ is the tangent space of $f(\Gamma)$ at $z = f(x)$, and if there is some possibility of confusion (note that the matrix $\nabla f(x) \circ \mathrm{T}_x \Gamma$ itself may not represent the orthogonal projection), it should be understood as the image of the map $\nabla f(x) \circ \mathrm{T}_x \Gamma \in \mathrm{Hom}(\mathbb{R}^n; \mathbb{R}^n)$, which should be the tangent space of $f(\Gamma)$ at $f(x)$. $|\Lambda_k \nabla f(x) \circ \mathrm{T}_x \Gamma|$ is the Jacobian of the map. If one chooses an orthonormal basis $v_1, \ldots, v_k$ of $T_x\Gamma$, the k-dimensional measure of the parallelepiped formed by $\nabla f(x) \circ v_1, \ldots, \nabla f(x) \circ v_k$ is precisely the Jacobian. If one forms the $n \times k$ matrix A whose columns are these vectors, the Jacobian can be also computed by $\sqrt{\det(A^\top \circ A)}$. One can also compute $\det(A^\top \circ A)$ as the sum of squares of determinants of $k \times k$ submatrices of A. Anyway, this is nothing fancy but just a change of variable formula. With this in mind, it is reasonable to define a mapping of general varifold $V \in \mathbf{G}_k(U)$ by a smooth diffeomorphism $f : U \to U$, which is called the push-forward of a varifold, as follows. For $\phi \in C_c(\mathbf{G}_k(U))$, define

$$f_\sharp V(\phi) := \int_{\mathbf{G}_k(U)} \phi(f(x), \nabla f(x) \circ S) |\Lambda_k \nabla f(x) \circ S| \, dV(x, S). \tag{1.5}$$

Note that the integrand is a continuous function of (x, S) and $\nabla f(x) \circ S$ is a k-dimensional subspace since f is a diffeomorphism. In the case that $V = |\Gamma|$, this reduces to (1.4). A little thought also shows that $f : U \to U$ need not be a diffeomorphism, but (1.5) makes sense as long as f is C^1 and $f^{-1}(K)$ is compact for all compact $K \subset U$. When the subspace $\nabla f(x) \circ S$ has a dimension less than k, the corresponding Jacobian is 0, so with this understanding, the integrand is still a continuous function of (x, S). Now an even less regular f such as Lipschitz poses a problem since we then do not have continuous ∇f anymore. However, if V is rectifiable, (1.5) still makes sense even if f is Lipschitz. This is not directly used much in this book so I skip the proof. The basic idea is that, if rectifiable, the integration can be decomposed into that on k-dimensional C^1 surfaces Γ_j as in Definition 1.1 and then f restricted to Γ_j is differentiable $\mathcal{H}^k$ a.e. there. This gives a measurable integrand and eventually a well-defined push-forward $f_\sharp V$ when V is rectifiable. Since the integrand in this case is not a continuous function of (x, S), on the other hand, the push-forward of a Lipschitz map does not behave continuously with respect to varifold convergence. This can be further remedied by considering a piece-wise smooth Lipschitz map as was done in [4], but this is also not needed subsequently.

1.5 The First Variation of a Varifold

Suppose that we are given a nice subset of varifolds such as rectifiable, integral, or unit density varifolds, with some control of their measures. To obtain a good compactness theorem for such a class, we typically need some extra information on how tangent spaces behave. A nice quantity for such a purpose is the first variation of a varifold, which will be discussed in this section. We also define a generalized notion of a mean curvature vector for a varifold.

We first recall a few basic facts. For any vector field $g = (g_1, \cdots, g_n)^\top \in C_c^1(U; \mathbb{R}^n)$, the usual divergence theorem applied to a domain $\Omega \subset U$ with C^1 boundary $\partial\Omega$ gives

$$\int_\Omega \operatorname{div} g \, d\mathcal{L}^n = \int_{\partial\Omega} g \cdot \nu \, d\mathcal{H}^{n-1}. \tag{1.6}$$

Here $\operatorname{div} g = \sum_{i=1}^n (g_i)_{x_i}$, $(g_i)_{x_i} = \frac{\partial g_i}{\partial x_i}$ and ν is the outward pointing unit vector to $\partial\Omega$. To consider a divergence theorem restricted to a k-dimensional surface, we define the notion of divergence restricted to a k-dimensional subspace.

Definition 1.8 For $S \in \mathbf{G}(n, k)$ and $g = (g_1, \cdots, g_n)^\top \in C_c^1(U; \mathbb{R}^n)$, define

$$\operatorname{div}_S g := \sum_{i=1}^k v_i \cdot \nabla_{v_i} g.$$

Here $\{v_1, \cdots, v_k\}$ is an orthonormal basis of S and $\nabla_{v_i} g$ is the component-wise directional derivative of g in the direction v_i. The definition does not depend on the choice of the orthonormal basis of S.

We may also define

$$\operatorname{div}_S g \text{ as } S \cdot \nabla g$$

equally. To see this, note that $S = \sum_{j=1}^k v_j \otimes v_j$ and choose $v_{k+1}, \cdots, v_n$ as a complementary orthonormal basis so that $\{v_1, \cdots, v_n\}$ forms a basis of $\mathbb{R}^n$. Then, since the trace of matrix P is obtained by $\sum_{i=1}^n v_i \cdot P(v_i)$,

$$S \cdot \nabla g = \operatorname{tr}(S \circ \nabla g) = \sum_{i=1}^n v_i \cdot (\sum_{j=1}^k v_j \otimes v_j \circ \nabla g)(v_i) = \sum_{i=1}^k \nabla_{v_i} g \cdot v_i = \operatorname{div}_S g.$$
$$\tag{1.7}$$

In the context of varifolds, the divergence comes up naturally because of the following formula.

Lemma 1.9 For $A \in \operatorname{Hom}(\mathbb{R}^n; \mathbb{R}^n)$ and $S \in \mathbf{G}(n, k)$, we have

$$\frac{d}{dt}\Big|_{t=0} |\Lambda_k(I + tA) \circ S| = S \cdot A. \tag{1.8}$$

Proof To see that this is true, since the claim is invariant under orthogonal change of variables, we may assume that S is the orthogonal projection to the first k coordinates $S : \mathbb{R}^n \to \mathbb{R}^k \times \{0\}$ and that the orthonormal basis of S is formed by the standard unit vectors. Then the determinant of $k \times k$ matrix $((I + tA) \circ S)^\top \circ (I +$

$tA) \circ S$ has $1 + 2t \sum_{j=1}^{k} A_{jj} + O(t^2)$ and $d/dt|_{t=0}(1 + 2t \sum_{j=1}^{k} A_{jj} + O(t^2))^{1/2} = \sum_{j=1}^{k} A_{jj}$. Since $S \cdot A = \sum_{j=1}^{k} A_{jj}$, we obtain the result. $\square$

Given $g =. (g_1, \ldots, g_n)^\top \in C_c^1(U; \mathbb{R}^n)$ which we now think more like a velocity and a k-dimensional C^1 surface Γ in U, consider $f_t : U \to U$ defined by $f_t(x) := x + tg(x)$, which is a bijective C^1 map for sufficiently small $t \in (-\varepsilon, \varepsilon)$. In the following, we consider the rate of change of the k-dimensional measure when Γ is pushed by the velocity given by g. For this, choose $\tilde{U} \subset\subset U$ such that $f_t \llcorner_{U \setminus \tilde{U}}$ is the identity map on $U \setminus \tilde{U}$. Note that no change is happening outside $\tilde{U}$ so we do not worry about it. Since

$$\mathcal{H}^k(f_t(\Gamma) \cap \tilde{U}) = \int_{\Gamma \cap \tilde{U}} |\Lambda_k(I + t\nabla g(x)) \circ T_x \Gamma| \, d\mathcal{H}^k(x),$$

using the formulas (1.8) and (1.7), we obtain

$$\frac{d}{dt}\Big|_{t=0} \mathcal{H}^k(f_t(\Gamma) \cap \tilde{U}) = \int_{\Gamma} \frac{\partial}{\partial t}\Big|_{t=0} |\Lambda_k(I + t\nabla g(x)) \circ T_x \Gamma| \, d\mathcal{H}^k(x)$$

$$= \int_{\Gamma} T_x \Gamma \cdot \nabla g(x) \, d\mathcal{H}^k(x) = \int_{\Gamma} \operatorname{div}_{T_x \Gamma} g(x) \, d\mathcal{H}^k(x).$$
$$(1.9)$$

This is the first variation of the k-dimensional area of Γ when moved by the velocity g. We can compute the analogous quantity, namely the rate of change of weight measure, for general varifold V using the push-forward generated by g as follows. By (1.5),

$$\frac{d}{dt}\Big|_{t=0} \|(f_t)_\sharp V\|(\tilde{U}) = \frac{d}{dt}\Big|_{t=0} \int_{\mathbf{G}_k(\tilde{U})} |\Lambda_k(I + t\nabla g(x)) \circ S| \, dV(x, S)$$

$$= \int_{\mathbf{G}_k(U)} S \cdot \nabla g(x) \, dV(x, S).$$
$$(1.10)$$

Note that the above interchange of integration and differentiation can be rigorously justified, and the resulting integrand $S \cdot \nabla g(x)$ is a continuous function on $\mathbf{G}_k(U)$ with compact support. Because of this, the following is naturally defined.

Definition 1.10 For $V \in \mathbf{V}_k(U)$ and $g \in C_c^1(U; \mathbb{R}^n)$, we define the **first variation** of V with respect to g through (1.10), i.e. with $\tilde{U} \subset\subset U$ with spt $g \subset \tilde{U}$,

$$\delta V(g) := \frac{d}{dt}\Big|_{t=0} \|(f_t)_\sharp V\|(\tilde{U}) = \int_{\mathbf{G}_k(U)} S \cdot \nabla g(x) \, dV(x, S). \qquad (1.11)$$

For $V = |\Gamma|$ with a k-dimensional C^1 surface Γ, we have

$$\delta|\Gamma|(g) = \int_{\mathbf{G}_k(U)} S \cdot \nabla g(x)\, \mathrm{d}|\Gamma|(x, S) = \int_\Gamma \mathrm{T}_x \Gamma \cdot \nabla g(x)\, \mathrm{d}\mathcal{H}^k(x), \qquad (1.12)$$

which is what it should be.

Now suppose that we have a k-dimensional C^2 surface Γ in U with C^1 boundary $\partial\Gamma$. Then the following formula is well-known. For all $g \in C_c^1(U; \mathbb{R}^n)$,

$$\int_\Gamma \mathrm{div}_{\mathrm{T}_x \Gamma}\, g(x)\, \mathrm{d}\mathcal{H}^k(x) = -\int_\Gamma h(\Gamma, x) \cdot g(x)\, \mathrm{d}\mathcal{H}^k(x) + \int_{\partial\Gamma} g(x) \cdot \nu(x)\, \mathrm{d}\mathcal{H}^{k-1}(x) \tag{1.13}$$

where $h(\Gamma, x)$ is the mean curvature vector of Γ at x and ν is the unit outward pointing co-normal vector for $\partial\Gamma$. The interpretation of (1.13) compared to (1.6) is that the contribution of "non-flatness" of Γ is reflected in the term involving $h(\Gamma, x)$. We observe that the right-hand side does not involve derivatives of g, while the left-hand does. This is more like an integration by parts formula, and when we are facing such a situation, we may consider a weak notion of differentiation, so to speak. The left-hand side can be naturally defined as the first variation for a general varifold. To make sense of the analogous object of the right-hand side of (1.13), we next consider the following.

Definition 1.11 For $V \in \mathbf{V}_k(U)$, we say that the first variation of V is **locally bounded** if, for any $\tilde{U} \subset\subset U$, there exists a constant $c(\tilde{U})$ such that for all $g \in C_c^1(\tilde{U}; \mathbb{R}^n)$, we have

$$\delta V(g) \le c(\tilde{U}) \sup_{x \in \tilde{U}} |g(x)|. \tag{1.14}$$

If $V = |\Gamma|$ with C^2 surface Γ and C^1 boundary $\partial\Gamma$, (1.12) and (1.13) show that

$$\delta|\Gamma|(g) \le \left(\int_{\Gamma \cap \tilde{U}} |h(\Gamma, x)|\, \mathrm{d}\mathcal{H}^k(x) + \mathcal{H}^{k-1}(\tilde{U} \cap \partial\Gamma) \right) \sup_{x \in \tilde{U}} |g(x)|. \tag{1.15}$$

Thus, this is an example of a locally bounded first variation since the quantities in the parenthesis are bounded for each $\tilde{U} \subset\subset U$. One can find examples of varifolds without locally bounded first variations, so the boundedness imposes some weak regularity on varifolds. In the following, we focus mostly on varifolds with locally bounded first variations.

Next, we proceed to define a weak notion of a mean curvature vector for a varifold with locally bounded first variation. We can proceed as follows.

Step 1. Due to (1.14), we may extend the domain of δV from $C_c^1(U; \mathbb{R}^n)$ to $C_c(U; \mathbb{R}^n)$ by approximation. Namely, given $g \in C_c(U; \mathbb{R}^n)$, we may choose an open set $\tilde{U} \subset\subset U$ such that $g \in C_c(\tilde{U}; \mathbb{R}^n)$ and a sequence $\{g^{(j)}\} \in C_c^1(\tilde{U}; \mathbb{R}^n)$ such that $\lim_{j \to \infty} \sup_{x \in \tilde{U}} |g^{(j)}(x) - g(x)| = 0$ by the standard approximation.

Then by (1.14), $\{\delta V(g^{(j)})\}$ is a Cauchy sequence in $\mathbb{R}$ and we define $\delta V(g) = \lim_{j\to\infty} \delta V(g^{(j)})$. One may check that $\delta V(g)$ does not depend on the choice of the approximating sequence, and that it satisfies the same inequality of (1.14) for all $g \in C_c(\tilde{U}; \mathbb{R}^n)$.

Step 2. Since δV is a locally bounded linear functional on $C_c(U; \mathbb{R}^n)$, the vector-version Riesz representation theorem [33] gives the existence of a Radon measure denoted by $\|\delta V\|$ and a $\|\delta V\|$ measurable function $\nu : U \to \mathbb{R}^n$ such that for all $g \in C_c(U; \mathbb{R}^n)$,

$$\delta V(g) = \int_U g(x) \cdot \nu(x) \, \mathrm{d}\|\delta V\|(x), \tag{1.16}$$

and $|\nu(x)| = 1$ for $\|\delta V\|$ a.e. $x \in U$.

Step 3. We apply the Radon–Nikodym theorem to the pair $\|\delta V\|$ and $\|V\|$. The absolutely continuous part of $\|\delta V\|$ is represented as a locally integrable and $\|V\|$ measurable function f. Writing the singular part as $\|\delta V\|_s$, we have

$$\mathrm{d}\|\delta V\| = f \, \mathrm{d}\|V\| + \mathrm{d}\|\delta V\|_s. \tag{1.17}$$

Of course, if $\|\delta V\| \ll \|V\|$, then $\mathrm{d}\|\delta V\|_s = 0$.

Step 4. Combining (1.16) and (1.17), we obtain

$$\delta V(g) = \int_U g(x) \cdot \nu(x) f(x) \, \mathrm{d}\|V\|(x) + \int_U g(x) \cdot \nu(x) \, \mathrm{d}\|\delta V\|_s(x). \tag{1.18}$$

In the case that $V = |\Gamma|$ with C^2 surface Γ and C^1 boundary $\partial\Gamma$, recall that $\|V\| = \mathcal{H}^k \llcorner \Gamma$. Also $\mathcal{H}^{k-1} \llcorner \partial\Gamma$ is a singular measure with respect to $\mathcal{H}^k \llcorner \Gamma$ since $\mathcal{H}^k(\partial\Gamma) = 0$. Thus, comparing (1.18) and (1.13), we naturally define the following.

Definition 1.12 Suppose $V \in \mathbf{V}_k(U)$ has a locally bounded first variation and define

$$h(V, x) := -\nu(x) f(x),$$

where $\nu(x)$ and $f(x)$ are obtained from V as above. The vector field $h(V, x)$ defined for $\|V\|$ a.e. $x \in U$ is called the **generalized mean curvature vector** of V.

Thus if $\|\delta V\| \ll \|V\|$, we have

$$\delta V(g) = -\int_U h(V, x) \cdot g(x) \, \mathrm{d}\|V\|(x) \tag{1.19}$$

for all $g \in C_c(U; \mathbb{R}^n)$. By the definition, $h(V, x)$ is a $\|V\|$ measurable locally integrable vector-valued function on U. It is also natural to consider $\|\delta V\|_s$ as a boundary measure and ν as a boundary co-normal for $\|\delta V\|_s$ a.e. in a generalized sense. In this book we mostly deal with the situation that $\|\delta V\|_s = 0$, that is, the case $\|\delta V\| \ll \|V\|$. A particularly important class of varifolds is the following.

Definition 1.13 $V \in \mathbf{V}_k(U)$ is called **stationary** if $\|\delta V\|(U) = 0$.

From the discussion above, a stationary varifold is a generalized k-dimensional surface which has no boundary and which has a vanishing mean curvature vector in U. More about stationary varifolds will be explained later.

For later use, here we define a notion of a weighted first variation of a general varifold. For any $V \in \mathbf{V}_k(U)$, $\phi \in C^1(U)$ and $g \in C_c^1(U; \mathbb{R}^n)$, we define

$$\delta(V, \phi)(g) := \int_{\mathbf{G}_k(U)} \phi(x) S \cdot \nabla g(x) + g(x) \cdot \nabla \phi(x) \, dV(x, S). \tag{1.20}$$

Using the same notation as before, this may also be obtained as the first variation of ϕV:

$$\frac{d}{dt}\bigg|_{t=0} \|(f_t)_\sharp(\phi V)\|(\tilde{U}) = \int_{\mathbf{G}_k(U)} \frac{\partial}{\partial t}\bigg|_{t=0} \phi(f_t(x))|\Lambda_k(I + t\nabla g(x)) \circ S| \, dV(x, S)$$

$$= \int_{\mathbf{G}_k(U)} \phi(x) S \cdot \nabla g(x) + g(x) \cdot \nabla \phi(x) \, dV(x, S).$$

If $\phi = 1$, then $\delta(V, 1)(g) = \delta V(g)$ as seen from (1.11). If $V = |\Gamma|$ for some smooth k-dimensional surface, then $S \cdot g(x) = \mathrm{div}_{\mathrm{T}_x \Gamma} g$ in the integral, thus

$$\delta(|\Gamma|, \phi)(g) = \int_\Gamma \phi(x) \mathrm{div}_{\mathrm{T}_x \Gamma} g + g(x) \cdot \nabla \phi(x) \, d\mathcal{H}^k(x).$$

Using $\phi \nabla g = \nabla(\phi g) - g \otimes \nabla \phi$ in (1.20), we have

$$\delta(V, \phi)(g) = \delta V(\phi g) + \int_{\mathbf{G}_k(U)} -S \cdot (g \otimes \nabla \phi) + g \cdot \nabla \phi \, dV$$

$$= \delta V(\phi g) + \int_{\mathbf{G}_k(U)} g \cdot (\nabla \phi - S(\nabla \phi)) \, dV \tag{1.21}$$

$$= \delta V(\phi g) + \int_{\mathbf{G}_k(U)} g \cdot S^\perp(\nabla \phi) \, dV$$

since $\nabla \phi - S(\nabla \phi) = S^\perp(\nabla \phi)$. If $V = |\Gamma|$, where Γ is a C^2 surface without boundary, and g is a normal vector field on Γ, we have

$$\delta(|\Gamma|, \phi)(g) = \int_\Gamma -\phi \, g \cdot h(\Gamma, \cdot) + g \cdot (\mathrm{T}_x \Gamma)^\perp(\nabla \phi) \, d\mathcal{H}^k$$

$$= \int_\Gamma (-\phi h(\Gamma, \cdot) + \nabla \phi) \cdot g \, d\mathcal{H}^k \tag{1.22}$$

due to (1.7) and (1.13).

1.6 Examples

Here we see some simple examples to better understand the abstract concepts in the preceding sections.

Define

$$\Gamma := \{(x_1, 0)^\top \in \mathbb{R}^2 : x_1 \geq 0\}.$$

Γ is a countably 1-rectifiable set since $\{(x_1, 0)^\top : x_1 \in \mathbb{R}\}$ is a C^1 1-dimensional curve and Γ is included in it. Let us denote

$$P_0 := \{(x_1, 0)^\top \in \mathbb{R}^2 : x_1 \in \mathbb{R}\} \in \mathbf{G}(2, 1).$$

Recall that we may identify P_0 with the 2×2 matrix representing the orthogonal projection to P_0. Except for the origin, which has a null $\mathcal{H}^1$ measure, we have P_0 as the usual tangent space on Γ. Define $|\Gamma| \in \mathbf{V}_1(\mathbb{R}^2)$ as described before. This means that for $\phi \in C_c(\mathbf{G}_1(\mathbb{R}^2))$, we have

$$\int_{\mathbf{G}_1(\mathbb{R}^2)} \phi(x, S)\, \mathrm{d}|\Gamma|(x, S) = \int_\Gamma \phi(x, T_x \Gamma)\, \mathrm{d}\mathcal{H}^1(x)$$

$$= \int_0^\infty \phi((x_1, 0)^\top, P_0)\, \mathrm{d}\mathcal{L}^1(x_1).$$

Let us see the first variation of $|\Gamma|$. For $g = (g_1, g_2)^\top \in C_c^1(\mathbb{R}^2; \mathbb{R}^2)$,

$$\nabla g = \begin{pmatrix} (g_1)_{x_1} & (g_1)_{x_2} \\ (g_2)_{x_1} & (g_2)_{x_2} \end{pmatrix},$$

$$P_0 \circ \nabla g = \begin{pmatrix} 1 & 0 \\ 0 & 0 \end{pmatrix} \circ \nabla g = \begin{pmatrix} (g_1)_{x_1} & (g_1)_{x_2} \\ 0 & 0 \end{pmatrix}$$

and

$$P_0 \cdot \nabla g = (g_1)_{x_1}.$$

Hence we have

$$\delta|\Gamma|(g) = \int_{\mathbf{G}_1(\mathbb{R}^2)} S \cdot \nabla g\, \mathrm{d}|\Gamma|(x, S) = \int_\Gamma P_0 \cdot \nabla g\, \mathrm{d}\mathcal{H}^1 = \int_0^\infty (g_1)_{x_1}(x_1, 0)\, \mathrm{d}x_1.$$

By integration by parts and since g has a compact support, we have

$$\delta|\Gamma|(g) = -g_1(0, 0).$$

In particular, we have $\delta|\Gamma|(g) \leq \sup|g|$, hence $|\Gamma|$ has a bounded first variation. We have $\||\Gamma|\| = \mathcal{H}^1 \llcorner_\Gamma$, while the support of $\delta|\Gamma|$ is concentrated at the origin, which has a null $\mathcal{H}^1$ measure. Thus $\delta|\Gamma|$ is singular with respect to $\||\Gamma|\|$, and the Radon–Nikodym theorem gives $f = 0$, $\nu = -(1, 0)^\top$; $\|\delta|\Gamma|\|_s$ is the delta measure at the origin, here denoted as m_0, so $\delta|\Gamma| = -(1, 0)^\top m_0$. In this case, $h(|\Gamma|, x) = 0$.

The argument of the previous paragraph can be extended to any half line rotated by some angle. For $v \in \mathbb{R}^2$ with $|v| = 1$, let $\Gamma_v := \{sv \in \mathbb{R}^2 : s \geq 0\}$. Since quantities under discussion are all coordinate free, we have $\delta|\Gamma_v| = -vm_0$, just as the $v = (1, 0)^\top$ case was $-(1, 0)^\top m_0$. Using this fact, we can form simple examples of stationary singular varifolds. Choose a set of unit vectors $v_1, \cdots, v_N$ so that $\sum_{i=1}^N v_i = 0$, and define $\Gamma := \cup_{i=1}^N \Gamma_{v_i}$. By the linearity of the definition of the first variation, we have

$$\delta|\Gamma| = \sum_{i=1}^N \delta|\Gamma_{v_i}| = -(\sum_{i=1}^N v_i)m_0 = 0.$$

Thus, for example, a varifold composed of three lines emanating from the origin and making equal angles of $120°$ is an example of a stationary 1-varifold which is also unit density.

As another simple example, let us consider $V \in \mathbf{RV}_k(U)$ such that $V = \theta|S|$, $S \in \mathbf{G}(n, k)$.

Proposition 1.14 $V = \theta|S|$ *is stationary if and only if θ is a constant.*

Proof Assume V is stationary. We may assume that $S = \mathbb{R}^k \times \{0\}$ after a suitable change of variables. Then for $g \in C_c^1(U; \mathbb{R}^n)$,

$$0 = \delta V(g) = \int_S (S \cdot \nabla g(x))\, \theta(x)\, \mathrm{d}\mathcal{H}^k(x) = \int_S \sum_{i=1}^k \frac{\partial g_i}{\partial x_i}(x)\theta(x)\, \mathrm{d}\mathcal{H}^k(x).$$

This means that the integrable function θ has $\nabla\theta = 0$ in a distributional sense, and this proves that θ is constant. If θ is constant, it is stationary. $\qquad\square$

1.7 Some Additional Properties of Integral Varifolds

One of the most important properties of integral varifolds is the following compactness theorem due to Allard [1].

Theorem 1.15 *Suppose $\{V^i\}_{i=1}^\infty \subset \mathbf{IV}_k(U)$ is a sequence of integral varifolds with*

$$\liminf_{i\to\infty}(\|V^i\|(K) + \|\delta V^i\|(K)) < \infty$$

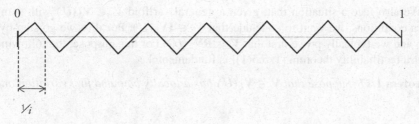

Fig. 1.1 Zig-zag shape converging to a line

for any compact $K \subset U$. Then there exist a subsequence $\{V^{i_j}\}_{j=1}^{\infty}$ and $V \in \mathbf{IV}_k(U)$
such that $V = \lim_{j \to \infty} V^{i_j}$ as measures on $\mathbf{G}_k(U)$.

Here, the point is not so much about the existence of a converging subsequence, but about the fact that the limit V is *integral*. Due to the general weak compactness theorem for Radon measures, by the local uniform bound on $\|V^i\|$, there exists a subsequence which converges to some general varifold $V \in \mathbf{V}_k(U)$. When we have also a local uniform bound on $\|\delta V^i\|$, the claim is that V is additionally integral. Intuitively, the bound on the first variations prevents wild "turning-around" of surfaces, in a certain weak sense. Let us consider an interesting example which exhibits the non-integral limit if the first variations are not bounded. Consider $V^i := |\Gamma_i| \in \mathbf{IV}_1(\mathbb{R}^2)$ which is a zig-zag curve as described in Fig. 1.1. Each zig-zag has length $\sqrt{2}/i$, and in any $B_R \subset \mathbb{R}^2$, there are at most $O(i)$ of them. Thus the weight measures $\|V^i\|(B_R)$ are locally bounded. On the other hand, V^i has a definite amount of first variation at each corner (due to the same computations above), thus locally, $\|\delta V^i\|(B_R) = O(i) \to \infty$ as $i \to \infty$. As $i \to \infty$, the zig-zag becomes smaller and smaller, and the support of $\|V^i\|$ approaches to $\mathbb{R} \times \{0\}$. One can check that $\|V^i\| \to \sqrt{2}\mathcal{H}^1 \llcorner_{\mathbb{R} \times \{0\}}$ as measures. As varifolds, V^i converges to some varifold V by the weak compactness theorem of measures, and thus $\|V^i\|$ converges to $\|V\|$ as well. But then $\|V\| = \sqrt{2}\mathcal{H}^1 \llcorner_{\mathbb{R} \times \{0\}}$ and V is not integral since $\sqrt{2} \notin \mathbb{N}$. It is a good exercise to check that V is not even rectifiable.

Related to the convergence of varifolds, it is good to note the following.

Proposition 1.16 *Suppose that a sequence $\{V^i\}_{i=1}^{\infty} \subset \mathbf{V}_k(U)$ converges to $V \in \mathbf{V}_k(U)$ as varifold. Then for any $g \in C_c^1(U; \mathbb{R}^n)$, $\lim_{i \to \infty} \delta V^i(g) = \delta V(g)$.*

This follows from the definition of δV and the varifold convergence (recall that $S \cdot \nabla g(x) \in C_c(\mathbf{G}_k(U))$). Thus, if we assume in addition in Theorem 1.15 that V^i is stationary ($\delta V^i = 0$), then V is a stationary integral varifold. It is a neat property that the class of stationary integral varifolds under uniform weight measure bound is compact under varifold convergence.

We also face a situation that, given a general varifold $V \in \mathbf{V}_k(U)$ with some extra conditions, we want to conclude that $V \in \mathbf{IV}_k(U)$. For this, we go step-by-step and we typically prove first that $V \in \mathbf{RV}_k(U)$. For this purpose, the following Allard rectifiability theorem [1, 5.5(1)] is fundamental.

Theorem 1.17 *Suppose that $V \in \mathbf{V}_k(U)$ has a locally bounded first variation and that*

$$\theta^{*k}(\|V\|, x) := \limsup_{r \to 0} \frac{1}{\omega_k r^k} \|V\|(B_r(x)) > 0 \qquad (1.23)$$

for $\|V\|$ a.e. $x \in U$. Then $V \in \mathbf{RV}_k(U)$.

This theorem tells us that if there is not too much wild turning-around (bounded first variation) and the weight measure is not spread-out (lower density bound), then V is rectifiable. The rectifiability tells us that there exist a $\mathcal{H}^k$ measurable countably k-rectifiable set Γ and a $\mathcal{H}^k$ measurable function θ such that $V = \theta|\Gamma|$. θ coincides with θ^{*k} for $\|V\|$ a.e. x. If we can prove that $\theta(x) \in \mathbb{N}$ for $\|V\|$ a.e. x, then we can conclude $V \in \mathbf{IV}_k(U)$. This last step requires some hard analysis and it is an interesting part.

The next property is something one does not even think about when dealing with smooth k-dimensional surfaces. Let $V \in \mathbf{V}_k(U)$ be a k-varifold with locally bounded first variation. Then as we discussed, δV defines a vector-valued measure and its absolutely continuous part with respect to $\|V\|$ defines a generalized mean curvature vector $h(V, x)$. If $V = |\Gamma|$ with smooth Γ with or without boundary, $h(V, x)$ coincides with the classical mean curvature vector on Γ. In such a smooth case, it is known that the mean curvature vector is perpendicular to the tangent space $T_x \Gamma$. In fact, it is often the case that the mean curvature vector is *defined* as the sum of principal curvatures multiplied by the unit normal vector with an appropriate choice of direction. On the other hand, it is not immediately clear if and how such property may hold for a general varifold, at least by simply examining the definition of $h(V, x)$ for a general V. Thus it is a little surprising that we have the following general result, due to Brakke [7, Ch. 5].

Theorem 1.18 *Suppose $V = \theta|\Gamma| \in \mathbf{IV}_k(U)$ with locally bounded first variation. Then for $\mathcal{H}^k$ a.e. $x \in \Gamma$, we have $h(V, x) \perp T_x \Gamma$. Here $T_x \Gamma$ is the approximate tangent space at x of Γ which exists $\mathcal{H}^k$ a.e. on Γ.*

We remark that one may not have such a property for non-integral varifolds: consider $V \in \mathbf{V}_1(\mathbb{R}^2)$ defined by $V := \theta|\{(x_1, 0)^\top : x_1 \in \mathbb{R}\}|$ with $\theta = (x_1)^2 + 1$. Then for $g = (g_1, g_2)^\top \in C_c^1(\mathbb{R}^2; \mathbb{R}^2)$,

$$\delta V(g) = \int_{\mathbb{R}} (g_1)_{x_1}((x_1)^2 + 1) \, dx_1 = -\int_{\mathbb{R}} 2x_1 g_1 \, dx_1$$

$$= -\int_{\mathbb{R}^2} \left(\frac{2x_1}{(x_1)^2 + 1}, 0\right)^\top \cdot g \, d\|V\|,$$

thus $h(V, x) = (\frac{2x_1}{(x_1)^2+1}, 0)^\top$, which is parallel to the tangent space of $\{(x_1, 0)^\top :$ $x_1 \in \mathbb{R}\}$.

The claim "$\mathcal{H}^k$ a.e. $x \in \Gamma$" is equivalent to "$\|V\|$ a.e. U" since $\|V\| = \theta \mathcal{H}^k \llcorner_\Gamma$ with $\theta \in \mathbb{N}$. Also $h(V, x) \perp \mathrm{T}_x \Gamma$ may be equally thought of as $S(h(V, x)) = 0$ for $(x, S) \in \mathbf{G}_k(U)$, V a.e. since $S(h(V, x)) = (\mathrm{T}_x \Gamma)(h(V, x))$.

Chapter 2
Definition of the Brakke Flow

2.1 Weak Formulation of Velocity

Suppose that we have a family of smooth surfaces $\{\Gamma(t)\}_{t\in[0,T)}$ for some $T \leq \infty$ with no boundaries in $U \subset \mathbb{R}^n$. Let $v = v(\Gamma(t), x)$ be the normal velocity vector of $\Gamma(t)$ at $x \in \Gamma(t)$. Here we consider how one may characterize the normal velocity using integration. The reason for such a pursuit is that, in the end, we want to replace $\Gamma(t)$ by a general varifold. To do so, let $\phi \in C_c^1(U \times [0, T); \mathbb{R}^+)$ be a non-negative "test function". We aim to characterize v in terms of the rate of change of

$$\int_{\Gamma(t)} \phi(x, t) \, d\mathcal{H}^k(x).$$

This can be directly described using the weighted first variation. Here, on the other hand, we use a parametric description which results in the same conclusion.

Let $\Phi_t(z) : \tilde{U} \subset \mathbb{R}^k \to U$ be a local parametrization of $\Gamma(t)$ so that $\operatorname{spt}\phi \cap \Gamma(t) \subset \Phi_t(\tilde{U})$ for $t \in (t_0 - \epsilon, t_0 + \epsilon)$. Then, we have

$$\left(\frac{\partial \Phi_t}{\partial t}(z)\right)^{\perp} = v(\Gamma(t), \Phi_t(z)),$$

where $\perp$ denotes the orthogonal projection to $(T_{\Phi_t(z)}\Gamma(t))^{\perp}$. Instead, it is convenient to have

$$\frac{\partial \Phi_t}{\partial t}(z)\bigg|_{t=t_0} = v(\Gamma(t_0), \Phi_{t_0}(z)) \tag{2.1}$$

on $\tilde{U}$ in the following. This can be achieved by introducing a local change of variables $z = z(\tilde{z}, t) = (z_1(\tilde{z}, t), \ldots, z_k(\tilde{z}, t))^{\top}$ which solves a suitable ODE

© The Author(s), under exclusive license to Springer Nature Singapore Pte Ltd. 2019
Y. Tonegawa, *Brakke's Mean Curvature Flow*, SpringerBriefs in Mathematics,
https://doi.org/10.1007/978-981-13-7075-5_2

system so that the tangential components of $\partial \Phi_t / \partial t$ vanish. In fact, to achieve this, $z = z(\tilde{z}, t)$ and $\tilde{\Phi}_t(\tilde{z}) = \Phi_t(z(\tilde{z}, t))$ should satisfy for each $j = 1, \ldots, k$

$$0 = \frac{\partial \tilde{\Phi}_t}{\partial t}(\tilde{z}) \cdot \frac{\partial \Phi_t}{\partial z_j}(z(\tilde{z}, t)) = \sum_{i=1}^{k} \frac{\partial \Phi_t}{\partial z_i} \cdot \frac{\partial \Phi_t}{\partial z_j} \frac{\partial z_i}{\partial t} + \frac{\partial \Phi_t}{\partial t} \cdot \frac{\partial \Phi_t}{\partial z_j},$$

where the right-hand side is evaluated at $z(\tilde{z}, t)$. With $z(\tilde{z}, t_0) = \tilde{z}$ as the initial data, one can solve this system for $z = z(\tilde{z}, t)$ for a short time. Thus assume in the following that (2.1) is satisfied. By the change of variables $x = \Phi_t(z)$, we have

$$\frac{\mathrm{d}}{\mathrm{d}t} \int_{\Gamma(t)} \phi(x, t) \, \mathrm{d}\mathcal{H}^k(x)$$
$$= \int_{\tilde{U}} \frac{\partial}{\partial t} \left\{ \phi(\Phi_t(z), t)(\det(\nabla \Phi_t(z)^\top \circ \nabla \Phi_t(z)))^{\frac{1}{2}} \right\} \mathrm{d}\mathcal{L}^k(z). \tag{2.2}$$

To compute the t derivative, fix any point $z_0 \in \tilde{U}$ and $x_0 = \Phi_{t_0}(z_0)$. After suitable rotation and parallel translation, we may assume that $z_0 = 0$, $t_0 = 0$ and $\nabla \Phi_0(0)$ is the inclusion $\mathbb{R}^k \to \mathbb{R}^k \times \{0\} = T_0 \Gamma(0) \subset \mathbb{R}^n$. We have

$$\Phi_t(z) = \Phi_0(z) + tv(\Gamma(0), \Phi_0(z)) + O(t^2)$$

and

$$\nabla \Phi_t(0) = \nabla \Phi_0(0) + t \nabla v(\Gamma(0), \Phi_0(0)) + O(t^2).$$

A little computation shows $\det(\nabla \Phi_t(0)^\top \circ \nabla \Phi_t(0)) = 1 + 2t \sum_{i=1}^{k} v_{z_i}^i + O(t^2)$ and we have

$$\frac{\partial}{\partial t} (\det(\nabla \Phi_t(z)^\top \circ \nabla \Phi_t(z)))^{\frac{1}{2}} \Big|_{z=0, t=0} = \sum_{i=1}^{k} v_{z_i}^i = \mathrm{div}_{T_{x_0} \Gamma(0)} v, \tag{2.3}$$

which is what we have seen before. Thus we obtain from (2.2) and (2.3) that

$$\frac{\mathrm{d}}{\mathrm{d}t} \int_{\Gamma(t)} \phi \, \mathrm{d}\mathcal{H}^k \Big|_{t=t_0} = \int_{\Gamma(t_0)} \nabla \phi \cdot v + \frac{\partial \phi}{\partial t} + \phi \, \mathrm{div}_{T_x \Gamma(t_0)} v \, \mathrm{d}\mathcal{H}^k. \tag{2.4}$$

Since v is the normal velocity, we have

$$\phi \, \mathrm{div}_{T_x \Gamma(t)} v = \mathrm{div}_{T_x \Gamma(t)} (\phi v) - \sum_{i=1}^{k} e_i \cdot (\nabla_{e_i} \phi) v = \mathrm{div}_{T_x \Gamma(t)} (\phi v)$$

using $e_i \cdot v = 0$, where $\{e_i\}_{i=1}^k$ is an orthonormal basis of $T_x \Gamma(t)$. Using this, (2.4) and (1.13), we obtain

$$\frac{d}{dt} \int_{\Gamma(t)} \phi \, d\mathcal{H}^k = \int_{\Gamma(t)} \nabla\phi \cdot v + \frac{\partial\phi}{\partial t} - \phi v \cdot h \, d\mathcal{H}^k$$

$$= \int_{\Gamma(t)} (\nabla\phi - \phi h) \cdot v + \frac{\partial\phi}{\partial t} \, d\mathcal{H}^k,$$

(2.5)

where $h = h(\Gamma(t), x)$ is the mean curvature vector of $\Gamma(t)$. Thus we verified that the normal velocity vector v satisfies (2.5) for any $\phi \in C_c^1(U \times [0, T); \mathbb{R}^+)$. Since it is an equality, we also have an inequality, $\leq$, in (2.5). The reader may find it rather strange to even talk about this, but let us continue. More about this point will be discussed in Sect. 2.3. We next see the converse statement. Suppose that we are given a vector field $\tilde{v}$ on $\Gamma(t)$ which is normal to $T_x\Gamma(t)$ at each x, t and which satisfies

$$\frac{d}{dt} \int_{\Gamma(t)} \phi \, d\mathcal{H}^k \leq \int_{\Gamma(t)} (\nabla\phi - \phi h) \cdot \tilde{v} + \frac{\partial\phi}{\partial t} \, d\mathcal{H}^k$$

(2.6)

for any $\phi \in C_c^1(U \times [0, T); \mathbb{R}^+)$. Then we claim that $\tilde{v}(x)$ has to coincide with the normal velocity vector $v(\Gamma(t), x)$. In fact, we have seen that the normal velocity vector satisfies (2.6) with an equality, thus by subtracting the equality we obtain

$$0 \leq \int_{\Gamma(t)} (\nabla\phi - \phi h) \cdot (\tilde{v}(x) - v(\Gamma(t), x)) \, d\mathcal{H}^k$$

(2.7)

for any $\phi \in C_c^1(U \times [0, T); \mathbb{R}^+)$, so in particular for any $\phi \in C_c^1(U; \mathbb{R}^+)$. Fix any point $x_0 \in \Gamma(t)$. Given any $\phi \in C_c^1(\mathbb{R}^n; \mathbb{R}^+)$ and $r > 0$, define $\phi_r(x) := r^{-k+1}\phi((x-x_0)/r)$. Substitute this ϕ_r into (2.7), change variables to $z = (x-x_0)/r$ and send $r \to 0+$. Note that the term involving h drops out since it is $O(r)$. Hence we obtain

$$0 \leq \int_{T_{x_0}\Gamma(t)} \nabla\phi(z) \, d\mathcal{H}^k(z) \cdot (\tilde{v}(x_0) - v(\Gamma_t, x_0)).$$

(2.8)

By integration by parts, $\int_{T_{x_0}\Gamma(t)} \nabla\phi(z) \, d\mathcal{H}^k(z) \in (T_{x_0}\Gamma(t))^\perp$, but otherwise, this can be an arbitrary vector by choosing an appropriate ϕ. Thus $\tilde{v}(x_0) = v(\Gamma(t), x_0)$ is proved, and since x_0 is arbitrary, they are equal on $\Gamma(t)$. The argument up to this point shows that (2.6) can be used to completely characterize the normal velocity vector if all the relevant quantities are smooth. Now going a step further, we obtain:

Proposition 2.1 *Suppose $\{\Gamma(t)\}_{t \in [0, T)}$ is a family of smooth k-dimensional surfaces without boundaries in $U \subset \mathbb{R}^n$. Then a smooth normal vector field v on $\Gamma(t)$*

is the normal velocity vector field if the following holds: for any $0 \leq t_1 < t_2 < T$
and $\phi \in C_c^1(U \times [0, T); \mathbb{R}^+)$, *we have*

$$\int_{\Gamma(t)} \phi(x, t) \, d\mathcal{H}^k \bigg|_{t=t_1}^{t_2} \leq \int_{t_1}^{t_2} \int_{\Gamma(t)} (\nabla \phi - \phi h) \cdot v + \frac{\partial \phi}{\partial t} \, d\mathcal{H}^k dt. \tag{2.9}$$

To get from (2.9) to (2.6), we divide both sides by $t_2 - t_1$ and let $t_2 \to t_1+$. If
a family of smooth surfaces happens to be a mean curvature flow, then (2.9) is
satisfied with $v = h$. As the above argument indicates, the justification of v being
the normal velocity requires relatively high regularity such as the differentiability of
the velocity v. So it is certainly not immediately clear if such a weak formulation is
useful. It is a surprise that it is indeed a good definition.

2.2 Weak Formulation of Normal Velocity for Varifolds and the Brakke Flow

Motivated by the idea of the previous section, we make a transition from smooth
surfaces to varifolds. Instead of going straight to the MCF, let us first define a
generalized notion of normal velocity for varifolds. To do this, it is reasonable
to replace $d\mathcal{H}^k \llcorner_{\Gamma(t)}$ by $d\|V_t\|$. It is also reasonable to stay close to the original
geometric problem that we require V_t to be integral for a.e. t, if not for all t. We
would like to consider the case of $v = h$, which results in having a $|h|^2$ term in
(2.9). Thus it is reasonable to ask for V_t to have a generalized mean curvature vector
$h(V_t, \cdot)$ in $L^2_{loc}(\|V_t\| \times dt)$ in the following. For simplicity, let us consider the case
where there are no boundaries, at least for a.e. t, so we also ask $\|\delta V_t\|_s = 0$ (or
$\|\delta V_t\| \ll \|V_t\|$) for a.e. t. Since it is a normal velocity, v should be perpendicular
to the approximate tangent space at least for a.e. sense. Under such a consideration,
we define:

Definition 2.2 Suppose a family $\{V_t\}_{t \in [0, T)} \subset \mathbf{V}_k(U)$ $(T \leq \infty)$ satisfies:

  (i) for a.e. t, $V_t \in \mathbf{IV}_k(U)$,
 (ii) for any compact $K \subset U$ and $t < T$, $\sup_{s \in [0, t]} \|V_s\|(K) < \infty$,
(iii) for a.e. t, V_t has locally bounded first variation and $\|\delta V_t\| \ll \|V_t\|$,
 (iv) $h(V_t, \cdot) \in L^2_{loc}(\|V_t\| \times dt)$.

A vector field $v \in L^2_{loc}(\|V_t\| \times dt)$ is called a generalized normal velocity for
$\{V_t\}_{t \in [0, T)}$ if the following conditions are satisfied:

  (v) for a.e. t and V_t a.e. $(x, S) \in \mathbf{G}_k(U)$, we have $S(v(x, t)) = 0$ and
 (vi) for all $0 \leq t_1 < t_2 < T$ and $\phi \in C_c^1(U \times [0, T); \mathbb{R}^+)$, we have

$$\int_U \phi(x, t)\, \mathrm{d}\|V_t\| \Big|_{t=t_1}^{t_2}$$

$$\leq \int_{t_1}^{t_2} \int_U (\nabla\phi(x, t) - \phi(x, t)h(V_t, x)) \cdot v(x, t) + \frac{\partial\phi}{\partial t}(x, t)\, \mathrm{d}\|V_t\|\mathrm{d}t.$$

$$(2.10)$$

In the definition, all the relevant measurabilities are assumed implicitly. For a smooth family of k-dimensional surfaces with no boundaries $\{\Gamma(t)\}_{t\in[0,T)}$, we may set $V_t = |\Gamma(t)|$. Then (i)–(iv) are satisfied, and v satisfying (v) and (vi) corresponds precisely to the classical normal velocity. Note that (v) is equivalent to $T_x\Gamma(t)(v(x, t)) = 0$ for $\mathcal{H}^k \llcorner_{\Gamma(t)}$ a.e. and for a.e. t due to (1.3). In this case, v is uniquely determined. Though I am not sure if there is any significance, in general, it is not clear to me if it is unique if such v exists. If (2.10) holds with equality for all $\phi \in C_c^1(U \times [0, T); \mathbb{R}^+)$, one can prove that it is unique, using an argument similar to the smooth case in the previous section.

As a special important class, we also define the following.

Definition 2.3 A family $\{V_t\}_{t\in[0,T)}$ is said to be **unit density** if V_t is unit density for a.e. t.

We finally define:

Definition 2.4 A family $\{V_t\}_{t\in[0,T)}$ is called a **Brakke flow** if (i)–(iv) of Definition 2.2 are satisfied and for all $0 \leq t_1 < t_2 < T$ and $\phi \in C_c^1(U \times [0, T); \mathbb{R}^+)$, we have

$$\int_U \phi(x, t)\, \mathrm{d}\|V_t\| \Big|_{t=t_1}^{t_2}$$

$$(2.11)$$

$$\leq \int_{t_1}^{t_2} \int_U (\nabla\phi(x, t) - \phi(x, t)h(V_t, x)) \cdot h(V_t, x) + \frac{\partial\phi}{\partial t}(x, t)\, \mathrm{d}\|V_t\|\mathrm{d}t,$$

that is, if we may take $v(x, t) = h(V_t, x)$ as a generalized normal velocity of $\{V_t\}_{t\in[0,T)}$.

Note that the condition (v) is automatically satisfied due to $V_t \in \mathbf{IV}_k(U)$ for a.e. t and Theorem 1.18. A smooth MCF $\{\Gamma(t)\}_{t\in[0,T)}$ defines a unit density Brakke flow by setting $V_t = |\Gamma(t)|$. On the other hand, a Brakke flow can potentially be, even with additional unit density assumption, non-smooth. Note that any stationary integral varifold is a time-independent Brakke flow since $\delta V = 0$ means $h = 0 = v$. Thus it is good to remember that Brakke flows are no smoother than stationary integral varifolds in general.

2.3   Remarks on the Definition

In characterizing the normal velocity, we replaced the equality by an inequality in (2.6). As a result of this, we inherited the same inequality in (2.10). Since the requirement of inequality is a relaxation of the definition, we may not wish to have this unless it is necessary. There are a few good reasons for this adoption, and somewhat unpleasant consequences as well. This is a complicated story, thus we start with a simple example which shows the necessity. ˙

Consider a 1-dimensional MCF as in Fig. 2.1. It consists of two circular curves of significantly different radii which are connected by a line segment. Since the smaller circle has larger curvature, it shrinks to a point while the other circle moves little.[1] At the moment the circle becomes a point, we have a line segment with a boundary point. As we saw in Sect. 1.6, we have a non-trivial singular measure $\|\delta V_t\|_s$ at the boundary point. The most natural way to continue the flow is to "remove" this line segment at the moment when this boundary appears. This operation creates a cusp on the circular curve, but the curvature flow immediately rounds off the cusp and becomes smooth. If we want to accommodate such instantaneous loss of some portion of curve, we cannot insist on having the equality in (2.10).

For the surface case, consider a disk whose boundary is attached to a thin torus as in Fig. 2.2. The torus has a relatively large principal curvature pointing towards the interior, so it is expected to shrink to a circle in a short time. At the moment that happens, we have a disk with boundary, and the only natural way to continue the flow is to eliminate the disk instantaneously.

Hence, when we study a Brakke flow in a general setting and not necessarily in the smooth setting, we cannot in general preclude such sudden loss of k-dimensional area, and thus we may not require the equality.

There is another justification for more technical and analytic reasons. In establishing the existence of a MCF, one often first constructs some approximate solutions. There are a few such methods. One is the original Brakke's existence

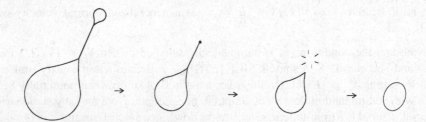

Fig. 2.1 Instantaneous vanishing of a line segment

---

[1]The behavior of this flow (resp. as well as that of the next torus example) is different from the Brakke flow established in [22] in that the line segment (resp. disk), which is considered as an "interior boundary" in the framework of [22], should vanish immediately instead.

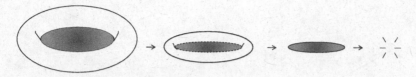

Fig. 2.2 Torus with the hole filled with a disk

theorem, where he first proves the existence of a time discrete approximate solution. Another is the so–called phase field approximation using the parabolic Allen–Cahn equation [18] or Ginzburg–Landau equation [6, 26]. There is also Ilmanen's elliptic regularization approach [19]. Here without going into the details, in all those cases, we have some approximate varifold V_t^ε and its approximate mean curvature $h_\varepsilon(V_t^\varepsilon, x)$ which satisfies (2.10) approximately. One hopes to obtain a solution by letting $\varepsilon \to 0$ and by making sure that V_t^ε converges to a nice integral varifold V_t. In doing so, while one can typically expect $\|V_t^\varepsilon\| \to \|V_t\|$, one can only expect a *weak convergence* of $h_\varepsilon(V_t^\varepsilon, x)$ to $h(V_t, x)$. Thus while the term involving $\nabla\phi \cdot h_\varepsilon(V_t^\varepsilon, x)$ may converge continuously, we can only prove the lower semicontinuous property for the quadratic term:

$$\int_U \phi|h(V_t, x)|^2 \, d\|V_t\| \leq \liminf_{\varepsilon \to 0+} \int_U \phi|h_\varepsilon(V_t^\varepsilon, x)|^2 \, d\|V_t^\varepsilon\|.$$

As a result of this, we can often prove only the inequality. In some special cases, it has been proved that we may have equality after all. Such is the case for the mean convex hypersurfaces [30]. The mean convexity is preserved under the MCF and the smallness of the dimension estimate of singularities due to [40, 41] contributes to the establishment of equality.

We mention some unpleasant aspect as the result of inequality. The inequality allows the following trivial non-uniqueness. Suppose that we have a Brakke flow $\{V_t\}_{t \in [0,T)}$. Then, for any $t_0 \in (0, T)$, define a family of varifolds $\{\tilde{V}_t\}_{t \in [0,T)}$ by setting $\tilde{V}_t = V_t$ for $t \in [0, t_0)$ and $\tilde{V}_t = 0$ for $t \geq t_0$. Then, since the inequality (2.10) allows a sudden loss of measure, one can check that $\{\tilde{V}_t\}_{t \geq 0}$ is a Brakke flow. Thus, given any initial varifold V_0, we always have a trivial solution $V_t = 0$ for all $t > 0$, which of course is not a desirable one to study. We should not allow such superfluous loss of measure in the definition of solution and one needs to supplement the solution with some extra properties. This supplementary part seems to have a certain degree of diversity at present and it strongly depends on the existence theory of the flow.

There are yet less obvious non-uniqueness issues, too, which may not involve any sudden loss of measure. For example, consider a set composed of two straight lines intersecting at a point with a right angle. This set itself has zero curvature, thus it is a time-independent solution. But one can also imagine a few possible solutions starting from this set; Fig. 2.3 gives the rough idea. In both of these solutions, there is no sudden loss of measure.

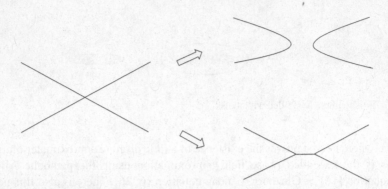

Fig. 2.3 Non-uniqueness without loss of measure

As for the definition of Brakke flow, the original definition in [7] is technically different from above in a few aspects. They are minor points, but just to avoid any confusion, let me explain here. In Brakke's original definition, he required

$$\overline{D_t}\|V_t\|(\phi) \leq \int_{\mathbf{G}_k(U)} (S^\perp(\nabla\phi) - \phi h) \cdot h + \frac{\partial\phi}{\partial t}\, dV_t \tag{2.12}$$

for all $\phi \in C_c^1(U \times [0, \infty); \mathbb{R}^+)$ and t in place of (2.10). Here $\overline{D_t}$ is the upper derivative. This formulation is more or less equivalent to (2.10) after integration and observing that h is perpendicular to the tangent space a.e. $\|V_t\|$ and a.e. t, hence $S^\perp(\nabla\phi) \cdot h$ in (2.12) can be replaced by $\nabla\phi \cdot h$. In this book, we work exclusively with the integral formulation instead of (2.12). What we found while generalizing the flow is that it is often convenient and natural to have the integral formulation of (2.10). When we generalize the flow to include an extra term, so that we have $v = h + u^\perp$, where u is a $\|V_t\| \times dt$ measurable vector field with some integrability condition and $u^\perp$ is the projection to the normal space, it is possible that (with $\cdot h$ replaced by $\cdot(h + u^\perp)$) both sides of (2.12) may be $+\infty$ and one needs to give a somewhat clumsy definition to deal with this problem. As far as the regularity theory of [21, 37] is concerned, one can work with the integral formulation for this more general flow. Moreover, there is a definite advantage when we establish the existence if one works with (2.10), as we saw in the proof of some existence theory in [35]. These are the reasons that we use the integral formulation (2.10) in place of (2.12). Another aspect of [7] is that Brakke also considered what we may call "rectifiable Brakke flow". The meaning of this is that, instead of requiring $V_t \in \mathbf{IV}_k(U)$, we may require less, and we ask that $V_t \in \mathbf{RV}_k(U)$ for a.e. t. In fact, if the initial varifold V_0 is only an element of $\mathbf{RV}_k(\mathbb{R}^n)$, Brakke outlined the existence of rectifiable Brakke flow in [7]. In this case, we lose the perpendicularity property of $h(V_t, \cdot)$ in general, as we saw in the example following Theorem 1.18. Thus, we would need to give a somewhat different formulation from (2.12) to take into account of the density variation of V_t. One reason that we do not pursue this direction is that we do not see how to establish the corresponding regularity theory for such rectifiable Brakke flow so far. In the regularity theory, the perpendicularity of the mean curvature vector is essential. In fact, even for the stationary case, the only known regularity theorem for non-constant density stationary varifolds is the Allard regularity theorem when the density function is sufficiently close to 1. In any case, the reason to only discuss the class of "integral Brakke flow" is that it is the most important class while general enough to include relevant geometric applications. So far, little is known about general rectifiable varifolds.

2.4 The Brakke Flow in General Riemannian Manifolds

For geometric applications, it is worthwhile to define a natural extension of the notion of Brakke flow when the ambient space is a general Riemannian manifold instead of $\mathbb{R}^n$. Suppose that we have a $\bar{k}$-dimensional Riemannian manifold $\mathcal{N}$ with $1 < \bar{k}$. By Nash's isometric imbedding theorem, we may consider $\mathcal{N}$ as a submanifold in $\mathbb{R}^n$, with the metric on $\mathcal{N}$ inherited from the ambient standard metric on $\mathbb{R}^n$. We want to consider a generalized notion of k-dimensional MCF in $\mathcal{N}$, where $1 \leq k < \bar{k} \leq n$. To do so, we first recall the second fundamental form of $\mathcal{N}$ at x denoted by $\mathbf{B}_x$. It is a bilinear form $\mathbf{B}_x : T_x\mathcal{N} \times T_x\mathcal{N} \longmapsto (T_x\mathcal{N})^\perp$ defined by

$$\mathbf{B}_x(v_1, v_2) := -\sum_{i=1}^{n-\bar{k}} (v_1 \cdot \nabla_{v_2} \tau_i) \tau_i \Big|_x, \quad v_1, v_2 \in T_x\mathcal{N}. \tag{2.13}$$

Here $\tau_1, \cdots, \tau_{n-\bar{k}}$ are locally defined vector fields which are orthonormal and which satisfy $\tau_i(y) \in (T_y\mathcal{N})^\perp$ on some neighborhood of x. Next, for $x \in \mathcal{N}$ and $S \in \mathbf{G}(n, k)$ with $S \subset T_x\mathcal{N}$, define

$$h_\mathcal{N}(x, S) := \sum_{i=1}^{k} \mathbf{B}_x(v_i, v_i) \in (T_x\mathcal{N})^\perp, \tag{2.14}$$

where $v_1, \cdots, v_k$ form an orthonormal basis of S. The normal vector field $h_\mathcal{N}(x, S)$ is independent of the choice of orthonormal basis of S. Let us consider a smooth k-dimensional submanifold $\Gamma \subset \mathcal{N}$. For any point $x \in \Gamma$, let $h(\Gamma, x)$ be the mean curvature vector of Γ as a submanifold of $\mathbb{R}^n$. Then we define

$$\tilde{h}(\Gamma, x) := h(\Gamma, x) - h_\mathcal{N}(x, T_x\Gamma). \tag{2.15}$$

The vector $\tilde{h}(\Gamma, x)$ is tangential to $\mathcal{N}$ and we have $(T_x\mathcal{N})(h(\Gamma, x)) = \tilde{h}(\Gamma, x)$. This can be seen by computing

$$\int_\Gamma (\tau_i \cdot h(\Gamma, x))\phi \, d\mathcal{H}^k = -\int_\Gamma \mathrm{div}_{T_x\Gamma}(\phi\tau_i) \, d\mathcal{H}^k = -\int_\Gamma \sum_{j=1}^k v_j \cdot \nabla_{v_j}(\phi\tau_i) \, d\mathcal{H}^k$$

$$= \int_\Gamma (h_\mathcal{N}(x, T_x\Gamma) \cdot \tau_i)\phi \, d\mathcal{H}^k$$

where τ_i is a locally defined vector field as in (2.13), $v_1, \cdots, v_k$ form a orthonormal basis of $T_x\Gamma$ and $\phi \in C_c^1(\mathbb{R}^n)$. We used (1.13), (2.13) and (2.14). Since ϕ is arbitrary, we have $\tau_i(x) \cdot h(\Gamma, x) = \tau_i(x) \cdot h_\mathcal{N}(x, T_x\Gamma)$, thus $(T_x\mathcal{N})^\perp(h(\Gamma, x)) = h_\mathcal{N}(x, T_x\Gamma)$. This tangential property also shows that, when we consider the first variation of area of Γ restricted within $\mathcal{N}$, the corresponding mean curvature vector is $\tilde{h}(\Gamma, x)$. Thus, a smooth family $\{\Gamma(t)\}_{t\in[0,T)}$ of k-dimensional submanifolds in $\mathcal{N}$ is naturally defined as the MCF in $\mathcal{N}$ when $v(\Gamma, x) = \tilde{h}(\Gamma, x)$. More generally, we may define the corresponding varifold version as follows.

Definition 2.5 Suppose that a family $\{V_t\}_{t\in[0,T)} \subset \mathbf{V}_k(\mathbb{R}^n)$ with spt $\|V_t\| \subset \mathcal{N}$ for all $t \in [0, T)$ satisfies (i)–(iv) of Definition 2.2. For a.e. $t \in [0, T)$ when $V_t \in \mathbf{IV}_k(\mathbb{R}^n)$ and $h(V_t, x) \in L_{loc}^2(\|V_t\|)$ exists, we define

$$\tilde{h}(V_t, x) := h(V_t, x) - h_\mathcal{N}(x, T_x\Gamma(t)),$$

where $T_x\Gamma(t)$ is the unique approximate tangent space of $\Gamma(t)$ at x and where $V_t = \theta|\Gamma(t)|$. We say $\{V_t\}_{t\in[0,T)}$ is a Brakke flow in $\mathcal{N}$ if:

(v') for all $0 \le t_1 < t_2 < T$ and $\phi \in C_c^1(\mathbb{R}^n \times [0, T); \mathbb{R}^+)$, we have

$$\int_\mathcal{N} \phi \, d\|V_t\|\Big|_{t=t_1}^{t_2} \le \int_{t_1}^{t_2} \int_\mathcal{N} (\nabla\phi - \phi\tilde{h}) \cdot \tilde{h} + \frac{\partial\phi}{\partial t} \, d\|V_t\|dt. \tag{2.16}$$

One can prove that $\tilde{h}(V_t, x) \in T_x\mathcal{N}$ for a.e. $t \in [0, T)$ and $\|V_t\|$ a.e. as a result of the assumptions (i)–(iv) and spt $\|V_t\| \subset \mathcal{N}$ (see [37, Section 7] for the details). The class of test functions $C_c^1(\mathbb{R}^n \times [0, T); \mathbb{R}^+)$ may be replaced by $C_c^1(\mathcal{N} \times [0, T); \mathbb{R}^+)$ since the derivatives in the direction of $(T_x\mathcal{N})^\perp$ do not affect the inequality at all. By these projections to $T_x\mathcal{N}$ in (2.16), the definition is intrinsic in $\mathcal{N}$. By definition, a smooth MCF in $\mathcal{N}$ is also a Brakke flow in $\mathcal{N}$. Analytically, one may consider this flow as a perturbation from the Brakke flow in $\mathbb{R}^n$ by a locally L^∞ bounded extra term $h_\mathcal{N}$, that is, $v = h - h_\mathcal{N}$. Since $h_\mathcal{N}$ depends on the approximate tangent space of V_t, we may not assume any further a priori regularity such as continuity on $h_\mathcal{N}$ for a general Brakke flow in $\mathcal{N}$.

Chapter 3
Basic Properties of the Brakke Flow

In this chapter, we present some basic properties satisfied by the Brakke flow.

3.1 Continuity Property of the Brakke Flow

First, let us prove two technical but frequently used lemmas.

Lemma 3.1 *For $\phi \in C_c^2(\mathbb{R}^n; \mathbb{R}^+)$, we have $\sup_{\{\phi>0\}} \frac{|\nabla\phi|^2}{\phi} \leq 2 \sup \|\nabla^2\phi\|$.*

Proof Fix an arbitrary point $\hat{x}$ such that $\phi(\hat{x}) > 0$. After suitable translation and orthogonal rotation, we may assume that $\hat{x} = 0$ and $|\nabla\phi(0)| = \phi_{x_1}(0)$. Since ϕ has a compact support, there exists a line segment I on the x_1 axis with the end point at the origin and the other end point at a, with the property that $\phi(a) = 0$ and $\phi(x) > 0$ in the interior of I. Since $\phi \geq 0$, $\phi(a) = 0$ implies $\phi_{x_1}(a) = 0$. By Cauchy's mean value theorem applied on I, there exists a point $b \in I$ different from a such that

$$\frac{\phi_{x_1}(0)^2}{\phi(0)} = \frac{2\phi_{x_1}(b)\phi_{x_1 x_1}(b)}{\phi_{x_1}(b)} = 2\phi_{x_1 x_1}(b).$$

This proves the claim. □

Lemma 3.2 *Given a Brakke flow as in Definition 2.4, for $\phi \in C_c^2(U; \mathbb{R}^+)$ and $t < T$, define*

$$f(s) := \|V_s\|(\phi) - C s,$$

© The Author(s), under exclusive license to Springer Nature Singapore Pte Ltd. 2019
Y. Tonegawa, *Brakke's Mean Curvature Flow*, SpringerBriefs in Mathematics,
https://doi.org/10.1007/978-981-13-7075-5_3

where

$$C = \sup_{U} \|\nabla^2 \phi\| \cdot \sup_{s \in [0,t]} \|V_s\|(\operatorname{spt}\phi).$$

Then f is a monotone decreasing function of s on $[0, t]$.

Proof By (2.11) and the Cauchy–Schwarz inequality

$$\nabla\phi \cdot h = \frac{\nabla\phi}{\sqrt{\phi}} \cdot \sqrt{\phi}\, h \leq \frac{1}{2}\frac{|\nabla\phi|^2}{\phi} + \frac{1}{2}\phi|h|^2$$

on $\{\phi > 0\}$, we have for any $0 \leq t_1 < t_2 \leq t$

$$\begin{aligned}
\|V_{t_2}\|(\phi) - \|V_{t_1}\|(\phi) &\leq \int_{t_1}^{t_2} \int_{\{\phi > 0\}} (\nabla\phi - \phi h) \cdot h \, \mathrm{d}\|V_s\| \mathrm{d}s \\
&\leq \int_{t_1}^{t_2} \int_{\{\phi > 0\}} -\frac{1}{2}\phi|h|^2 + \frac{1}{2}\frac{|\nabla\phi|^2}{\phi} \, \mathrm{d}\|V_s\| \mathrm{d}s.
\end{aligned} \tag{3.1}$$

By Lemma 3.1, the right-hand side is $\leq C(t_2 - t_1)$, which proves the claim. $\quad\square$

Using Lemma 3.2, we can show:

Proposition 3.3 *Given a Brakke flow as in Definition 2.4, there exists a set $B \subset [0, T)$ which is at most countable such that $\|V_t\|(\phi)$ is a continuous function of t on $[0, T) \setminus B$ for all $\phi \in C_c(U; \mathbb{R}^+)$.*

Proof We may choose a countable set $\{\phi_j\}_{j\in\mathbb{N}} \subset C_c^2(U; \mathbb{R}^+)$ such that it is dense in $C_c(U; \mathbb{R}^+)$ with respect to the sup-norm $\sup_U |\cdot|$. By Lemma 3.2, for each fixed j, $\|V_s\|(\phi_j) - Cs$ is a monotone decreasing function of s, thus it is continuous except for at most a countable subset of $[0, T)$. Let B be the union over j of such discontinuity sets corresponding to ϕ_j. By using the density of $\{\phi_j\}_{j\in\mathbb{N}}$, one may prove that $\|V_s\|(\phi)$ is continuous on $[0, T) \setminus B$ for each $\phi \in C_c(U; \mathbb{R}^+)$. $\quad\square$

3.2 Huisken's Monotonicity Formula

Assume for simplicity that we have a Brakke flow $\{V_t\}_{t\in[0,T)}$ in $\mathbb{R}^n$ all contained in a bounded domain, say, B_R for some $R > 0$. The following formula due to Huisken in the smooth case [17] is an extremely useful tool for the analysis of singularities and regularity theories for the Brakke flow. Let us define for $x, y \in \mathbb{R}^n$ and $s > t$ that

$$\rho_{(y,s)}(x, t) := \frac{1}{(4\pi(s - t))^{\frac{k}{2}}} \exp\left(-\frac{|x - y|^2}{4(s - t)}\right).$$

This is a time-reversed k-dimensional heat kernel with pole at (y, s). As $t \nearrow s$, $\rho_{(y,s)}(x, t)$ concentrates at y as a function of x.

Theorem 3.4 *For a Brakke flow* $\{V_t\}_{t \in [0,T)}$ *in* $\mathbb{R}^n$ *with* spt $\|V_t\| \subset B_R$, *we have*

$$\int_{\mathbb{R}^n} \rho_{(y,s)}(x, t) \mathrm{d}\|V_t\|(x) \Big|_{t=t_1}^{t_2}$$

$$\leq -\int_{t_1}^{t_2} \int_{\mathbf{G}_k(\mathbb{R}^n)} \left| h(V_t, x) + \frac{S^\perp(x - y)}{2(s - t)} \right|^2 \rho_{(y,s)}(x, t) \, \mathrm{d}V_t(x, S) \mathrm{d}t$$

for any $0 \leq t_1 < t_2 < s$.

In particular, $\int_{\mathbb{R}^n} \rho_{(y,s)}(x, t) \, \mathrm{d}\|V_t\|(x)$ is monotone decreasing for $t \in [0, s)$.

Proof Write $\rho = \rho(x, t) = \rho_{(y,s)}(x, t)$. A rather surprising direct computation for any $S \in \mathbf{G}(n, k)$ shows (recall $S = \sum_{j=1}^k v_j \otimes v_j$ where $\{v_1, \ldots, v_k\}$ is an orthonormal basis of S) that we have

$$\frac{\partial \rho}{\partial t} + S \cdot \nabla^2 \rho + \frac{|S^\perp(\nabla \rho)|^2}{\rho} = 0. \tag{3.2}$$

Let us check this for the case $S = \mathbb{R}^k \times \{0\}$. We have

$$\frac{\partial \rho}{\partial t} = \frac{k}{2(s - t)} \rho - \frac{|x - y|^2}{4(s - t)^2} \rho,$$

$$S \cdot \nabla^2 \rho = \sum_{i=1}^k \frac{\partial^2 \rho}{\partial x_i^2} = -\frac{k}{2(s - t)} \rho + \frac{\sum_{i=1}^k (x_i - y_i)^2}{4(s - t)^2} \rho,$$

$$\frac{|S^\perp(\nabla \rho)|^2}{\rho} = \frac{\sum_{i=k+1}^n (x_i - y_i)^2}{4(s - t)^2} \rho,$$

and they add up to be 0 indeed. For $S = \sum_{j=1}^k v_j \otimes v_j$, one can compute similarly. In (2.11), use $\phi(x, t) = \rho_{(y,s)}(x, t)$. ρ does not have a compact support, but since spt $\|V_t\| \subset B_R$, we can justify such use by a truncation outside B_R. Writing $h = h(V_t, x)$, we compute

$$\int_{\mathbb{R}^n} (\nabla \rho - \rho h) \cdot h \, \mathrm{d}\|V_t\| = \int_{\mathbb{R}^n} 2\nabla \rho \cdot h - \nabla \rho \cdot h - \rho |h|^2 \, \mathrm{d}\|V_t\|$$

$$= \int_{\mathbf{G}_k(\mathbb{R}^n)} 2S^\perp(\nabla \rho) \cdot h + S \cdot \nabla^2 \rho - \rho |h|^2 \, \mathrm{d}V_t. \tag{3.3}$$

Here, we used Theorem 1.18 to conclude $2\nabla\rho \cdot h = 2S^{\perp}(\nabla\rho) \cdot h$ for V_t a.e., and (1.11) and (1.19) to conclude

$$-\int_{\mathbb{R}^n} \nabla\rho \cdot h \, d\|V_t\| = \delta V_t(\nabla\rho) = \int_{G_k(\mathbb{R}^n)} S \cdot \nabla^2\rho \, dV_t.$$

For the two terms in (3.3), we next complete the square as

$$2S^{\perp}(\nabla\rho) \cdot h - \rho|h|^2 = -\left|h - \frac{S^{\perp}(\nabla\rho)}{\rho}\right|^2 \rho + \frac{|S^{\perp}(\nabla\rho)|^2}{\rho}.$$

Thus we have

$$\int_{\mathbb{R}^n} (\nabla\rho - \rho h) \cdot h \, d\|V_t\| = \int_{G_k(\mathbb{R}^n)} -\rho\left|h - \frac{S^{\perp}(\nabla\rho)}{\rho}\right|^2 + \frac{|S^{\perp}(\nabla\rho)|^2}{\rho} + S \cdot \nabla^2\rho \, dV_t.$$

(3.4)

Now, going back to (2.11) and using (3.4) and (3.2), we obtain for any $0 \le t_1 < t_2 < s$

$$\int_{\mathbb{R}^n} \rho \, d\|V_t\|\Big|_{t=t_1}^{t_2} \le \int_{t_1}^{t_2} \int_{\mathbb{R}^n} (\nabla\rho - \rho h) \cdot h + \frac{\partial\rho}{\partial t} \, d\|V_t\| dt$$

$$\le \int_{t_1}^{t_2} \int_{G_k(\mathbb{R}^n)} -\rho\left|h - \frac{S^{\perp}(\nabla\rho)}{\rho}\right|^2 \, dV_t \qquad (3.5)$$

$$\le 0.$$

Since $\nabla\rho/\rho = -(x - y)/2(s - t)$, this proves the desired inequality.   □

When we consider a Brakke flow on a bounded domain $U \subset \mathbb{R}^n$, it is straightforward to localize above. For example, without loss of generality, suppose that $B_3 \subset U$ and $y \in B_1$. Choose a suitable cut-off function $0 \le \zeta \le 1$ such that $\zeta = 1$ on B_2 and $= 0$ on $U \setminus B_3$. We can repeat the same computations for $\hat{\rho}_{(y,s)}(x, t) := \rho_{(y,s)}(x, t)\zeta(x)$ in place of $\rho_{(y,s)}(x, t)$ above. Then the computations of (3.2) for $\hat{\rho}$ result in extra terms coming from differentiation of ζ times ρ or $\nabla\rho$, which are non-zero only on $B_3 \setminus B_2$. For $y \in B_1$, $\rho_{(y,s)}(x, t)$ (and similarly its derivatives) is bounded by $c(k)(s - t)^{-k/2} \exp(-1/4(s - t))$ on $B_3 \setminus B_2$. This function is bounded by $c(k) \exp(-1/8(s - t))$ for all $t < s$. Thus we have a similar formula for $y \in B_1$ with such $\hat{\rho}$, namely, for any $0 \le t_1 < t_2 < s$,

$$\int_{B_3} \hat{\rho}_{(y,s)} \, d\|V_t\|\Big|_{t=t_1}^{t_2} \le -\int_{t_1}^{t_2} \int_{G_k(B_3)} \left|h + \frac{S^{\perp}(x - y)}{2(s - t)}\right|^2 \hat{\rho}_{(y,s)} \, dV_t dt$$

$$+ c(k) \sup_{t \in [t_1, t_2]} \|V_t\|(B_3) \int_{t_1}^{t_2} \exp(-1/8(s - t)) \, dt.$$

If we change variables, $\tilde{x} = xR$ and $\tilde{t} = tR^2$ so that $B_{3R} \subset U$ and $y \in B_R$, we have

$$\int_{B_{3R}} \hat{\rho}_{(y,s)} \, d\|V_t\| \Big|_{t=t_1}^{t_2} \le - \int_{t_1}^{t_2} \int_{\mathbf{G}_k(B_{3R})} \Big| h + \frac{S^\perp(x-y)}{2(s-t)} \Big|^2 \hat{\rho}_{(y,s)} \, dV_t dt$$

$$+ c(k) R^{-k-2} \sup_{t \in [t_1, t_2]} \|V_t\|(B_{3R}) \int_{t_1}^{t_2} \exp(-R^2/8(s-t)) \, dt.$$

(3.6)

An easy consequence of the above is the following estimate which may be called a uniform mass density ratio bound from above.

Proposition 3.5 *Suppose* $\{V_t\}_{t \in [0,T)}$ *is a Brakke flow in* U. *For any* $0 < \delta < T$ *and* $B_{3R}(x_0) \subset U$, *there exist a constant* c *depending only on* δ, k *and* R *with the following property. For any* $t \in [\delta, T)$ *and* $B_r(y) \subset B_R(x_0)$, *we have*

$$r^{-k} \|V_t\|(B_r(y)) \le c \sup_{s \in [0,t]} \|V_s\|(B_{3R}(x_0)).$$

Proof Suppose that $x_0 = 0$ without loss of generality. For such $B_r(y)$, we use (3.6) with $t_2 = t$, $t_1 = 0$ and $s = t + r^2$. Then we have

$$\int_{B_{3R}} \hat{\rho}_{(y,t+r^2)}(x,t) \, d\|V_t\| \le \int_{B_{3R}} \hat{\rho}_{(y,t+r^2)}(x,0) \, d\|V_0\| + c(R) \sup_{s \in [0,t]} \|V_s\|(B_{3R}).$$

Since $t \ge \delta$, $\hat{\rho}_{(y,t+r^2)}(x,0) \le \frac{1}{(4\pi\delta)^{\frac{k}{2}}}$. On $B_r(y)$, we have $\hat{\rho}_{(y,t+r^2)}(x,t) \ge \frac{1}{(4\pi r^2)^{\frac{k}{2}}} \exp\left(-\frac{1}{4}\right)$. Then, with a suitable choice of c, the claim follows.   □

This is a reassuring property of a Brakke flow in that the measures behave like a k-dimensional measure uniformly from above on all small scales. We can also derive the following.

Proposition 3.6 *Suppose that* $\{V_t\}_{t \in [0,T)}$ *is a Brakke flow in* U *and that* $r^{-k} \|V_t\|(B_r(x)) \le E_1$ *for all* $t \in (0,T)$ *and* $B_r(x) \subset U$. *Then there exist* $c > 0$ *and* $\gamma > 0$ *depending only on* k *and* E_1 *with the following property. Suppose that* $\|V_t\|(B_R(y)) > 0$, $B_{3R}(y) \subset U$ *and* $t > cR^2$. *Then*

$$\|V_{t-cR^2}\|(B_{3R}(y)) \ge \gamma R^k.$$

The idea of the claim is that, if there is some amount of MCF in $B_R(y)$ at time t, then, going back in time, there must be some definite amount of k-dimensional MCF in $B_{3R}(y)$. Note that the statement is parabolically scale-invariant. Conversely, if there is not enough measure at $t - cR^2$ in $B_{3R}(y)$, then there is no MCF in $B_R(y)$ at t. The smallness of measure allows very complicated surfaces, namely, we can have extremely long branching tentacles in $B_{3R}(y)$ with very small surface measure for $k = 2$. Under the MCF, such a surface has to vanish in $B_R(y)$ after the elapse of time of cR^2. A similar flavor of idea is expressed in Brakke's "clearing-out lemma" [7, 6.3]. The following proof is a little technical and can be skipped, but I include it for the interested reader.

Proof Recall that $V_t \in \mathbf{IV}_k(U)$ for a.e. t. We fix an arbitrary $t \in (0, T)$. Suppose first that $V_t \in \mathbf{IV}_k(U)$, that is, $V_t = \theta|M|$ for an $\mathcal{H}^k$-measurable countably k-rectifiable set M and an $\mathcal{H}^k$ a.e. integer-valued θ defined on M. We may assume that $y = 0$. Since $\|V_t\|(B_R) > 0$, there exists $y' \in B_R$ such that M has an approximate tangent space at y', $\theta(y') \in \mathbb{N}$, $\theta(y') = \lim_{r \to 0} (\omega_k r^k)^{-1} \int_{M \cap B_r(y')} \theta(x) \, d\mathcal{H}^k(x)$. The last three properties are satisfied $\mathcal{H}^k$ a.e. on M and that is why we have such a point in B_R. With somewhat cumbersome error estimates, we can also prove that, with $\hat{\rho}$ defined as above,

$$\lim_{\delta \to 0+} \int_{M \cap B_{3R}} \hat{\rho}_{(y', t+\delta^2)}(x, t) \, d\|V_t\|(x) = \theta(y') \int_{T_{y'}M} \rho_{(0,1)}(x, 0) \, d\mathcal{H}^k(x)$$

$$= \theta(y'),$$

which is ≥ 1. The idea is that $\hat{\rho}_{(y', t+\delta^2)}(x, t) \, d\|V_t\|$ after a change of variable $z = (x - y')/\delta$ is approximated by $\rho_{(0,1)}(z, 0)\theta(y') \, d\mathcal{H}^k \llcorner_{T_{y'}M}$ when δ is small. We then use (3.6) with $s = t + \delta^2$, $t_2 = t$ and $t_1 < t_2$ to be chosen and let $\delta \to 0+$. Then we have

$$1 \leq \theta(y') \leq \int_{B_{3R}} \hat{\rho}_{(y', t)}(x, t_1) \, d\|V_{t_1}\| + c(k)E_1 R^{-2} \int_{t_1}^{t} \exp(-R^2/8(t - s)) \, ds.$$

Fix a small $c > 0$ depending only on k and E_1 such that the last term is less than $1/2$ for $t_1 = t - cR^2$. With this t_1, $\hat{\rho}_{(y', t)}(x, t_1) \leq c(k)R^{-k}$ and we obtain the desired estimate for this t_1. If $V_t \notin \mathbf{IV}_k(U)$, we use the fact that there is an arbitrarily close time $t' < t$ such that $V_{t'} \in \mathbf{IV}_k(U)$. In fact, for all sufficiently close $t' < t$ and arbitrarily small $\epsilon > 0$, we have $\|V_{t'}\|(B_{R+\epsilon}) > 0$. This is because we have $\|V_t\|(\phi) \leq \|V_{t'}\|(\phi) + c(\phi)(t - t')$ for all $\phi \in C_c^2(U; \mathbb{R}^+)$ by Lemma 3.2 and if there exists $t_i \to t-$ such that $\|V_{t_i}\|(B_{R+\epsilon}) = 0$, then this shows that $\|V_t\|(B_R) = 0$, a contradiction. In particular, we have a sequence of $t_i \to t-$ such that $\|V_{t_i}\|(B_{R+\epsilon}) > 0$ and $V_{t_i} \in \mathbf{IV}_k(U)$. For this t_i, we may repeat the same argument (with $B_{R+\epsilon}$) and we may obtain the desired result. $\square$

3.3   Compactness Property for the Brakke Flow

It is important to know that the set of Brakke flows with a suitable bound has a good compactness property. Since this is a good place to apply what we know up to this point, I give a detailed proof for this part.

Theorem 3.7 *Suppose $\{V_t^i\}_{t \in [0,T), i \in \mathbb{N}}$ is a family of Brakke flows in $U \times [0, T)$ such that $\sup_{t \in [0,T), i \in \mathbb{N}} \|V_t^i\|(K) < \infty$ for any compact $K \subset U$. Then there exist a subsequence $\{i_j\}_{j \in \mathbb{N}}$ and a Brakke flow $\{V_t\}_{t \in [0,T)}$ such that for all $t \in [0, T)$, we have*

$$\lim_{j \to \infty} \|V_t^{i_j}\| = \|V_t\|$$

as Radon measures on U. Moreover, for a.e. $t \in [0, T)$, there exists a subsequence $\{i'_j\}_{j \in \mathbb{N}} \subset \{i_j\}_{j \in \mathbb{N}}$ (which may depend on t) such that

$$\lim_{j \to \infty} V_t^{i'_j} = V_t$$

as varifolds on U.

It is worthwhile to emphasize that we can choose once and for all a subsequence $\{i_j\}_{j \in \mathbb{N}}$ independent of t such that $\|V_t^{i_j}\|$ converges to $\|V_t\|$ for all $t \in [0, T)$, not just a.e. t. On the other hand, we do not know in general that the full subsequence $V_t^{i_j}$ converges to V_t in the varifold topology, i.e. as measures on $\mathbf{G}_k(U)$. For the varifold convergence, we can only choose for a.e. t a further subsequence which in general may depend on t.

Proof Let $\{\phi_k\}_{k \in \mathbb{N}} \subset C_c^2(U; \mathbb{R}^+)$ be a countable set of functions such that it is dense in $C_c(U; \mathbb{R}^+)$ with respect to the sup-norm $\sup_U | \cdot |$. As in Lemma 3.2, for each i and ϕ_k, $f_{i,k}(s) := \|V_s^i\|(\phi_k) - c(k)s$ is a monotone decreasing function of s on $[0, T)$. Here,

$$c(k) := \sup \|\nabla^2 \phi_k\| \cdot \sup_{t \in [0,T), i \in \mathbb{N}} \|V_t^i\|(\operatorname{spt} \phi_k)$$

is a constant which depends only on k. Take any dense countable set $A \subset [0, T)$ and fix it. Since $\{f_{i,k}(s)\}_{i \in \mathbb{N}}$ is a bounded sequence for each fixed k and s, a diagonal argument proves that there exists a subsequence $\{i_j\} \subset \mathbb{N}$ such that for each $k \in \mathbb{N}$ and $s \in A$, $\{f_{i_j,k}(s)\}_{j \in \mathbb{N}}$ converges to a number $g_k(s)$. As a function of $s \in A$, $g_k(s)$ is a monotone decreasing function on A, and one can extend $g_k(s)$ to $[0, T)$ from A so that the resulting function (denoted by $g_k(s)$ again) will be continuous from the right on $[0, T) \setminus A$. There are at most countably many discontinuity points for $g_k(s)$ on $[0, T)$, and by the monotone decreasing property, one can show that $\{f_{i_j,k}(s)\}_{j \in \mathbb{N}}$ converges to $g_k(s)$ at any continuity point of $g_k(s)$. This shows that $\{f_{i_j,k}(s)\}_{j \in \mathbb{N}}$ converges except for the union of discontinuity points of $g_k(s)$. Thus, except for a countable set of points in $[0, T)$, $\{\|V_s^{i_j}\|(\phi_k)\}_{j \in \mathbb{N}}$ converges for each $k \in \mathbb{N}$. Since such exceptional points are countable, we may carry out another diagonal argument so that the further subsequence (denoted by the same index) $\{\|V_s^{i^j}\|(\phi_k)\}_{j\in\mathbb{N}}$ converges for all $s \in [0, T)$ and each fixed $k \in \mathbb{N}$. By the density argument, for any fixed $\phi \in C_c(U)$, $\{\|V_s^{i^j}\|(\phi)\}_{j\in\mathbb{N}}$ is also convergent, and the limit defines a locally bounded linear functional on $C_c(U)$ for each $s \in [0, T)$. By the Riesz representation theorem applied to the limit functional, there exists a Radon measure denoted by μ_s on U, i.e. we have

$$\lim_{j\to\infty} \|V_s^{i^j}\|(\phi) = \mu_s(\phi)$$

for all $s \in [0, T)$ and $\phi \in C_c(U)$. This still does not show the existence of a limit varifold, and we examine it next. Because of (3.1), for $\phi \in C_c^2(U; \mathbb{R}^+)$ and j, we have

$$\frac{1}{2} \int_0^T \int_U \phi |h(V_s^{i^j}, \cdot)|^2 \, d\|V_s^{i^j}\| ds \le \frac{1}{2} \int_0^T \int_{\{\phi>0\}} \frac{|\nabla\phi|^2}{\phi} \, d\|V_s^{i^j}\| ds + \|V_0^{i^j}\|(\phi).$$

The right-hand side is uniformly bounded with respect to j. Thus, by Fatou's Lemma, for a.e. $s \in [0, T)$, we have

$$\liminf_{j\to\infty} \int_U \phi |h(V_s^{i^j}, \cdot)|^2 \, d\|V_s^{i^j}\| < \infty. \tag{3.7}$$

By choosing a suitable sequence of $C_c^2(U; \mathbb{R}^+)$ functions $\{\phi_k\}_{k\in\mathbb{N}}$ such that $\{\phi_k = 1\}$ monotonically increase to U, we may prove that for a.e. $s \in [0, T)$ and for all $\tilde{U} \subset\subset U$,

$$\liminf_{j\to\infty} \int_{\tilde{U}} |h(V_s^{i^j}, \cdot)|^2 \, d\|V_s^{i^j}\| < \infty. \tag{3.8}$$

In particular, since we have

$$\|\delta V\|(\tilde{U}) = \int_{\tilde{U}} |h(V, \cdot)| \, d\|V\| \le \left(\int_{\tilde{U}} |h(V, \cdot)|^2 \, d\|V\| \right)^{\frac{1}{2}} (\|V\|(\tilde{U}))^{\frac{1}{2}},$$

we may use Theorem 1.15 to a.e. $s \in [0, T)$ so that there exists a subsequence $\{i'_j\}_{j\in\mathbb{N}} \subset \{i_j\}_{j\in\mathbb{N}}$ and $V_s \in \mathbf{IV}_k(U)$ such that $\lim_{j\to\infty} V_s^{i'_j} = V_s$ as varifolds. Note that the choice of the subsequence may depend on s. On the other hand, we know that $\{\|V_s^{i'_j}\|\}_{j\in\mathbb{N}}$ converges to μ_s, so we have $\mu_s = \|V_s\|$. For integral varifold V_s, there exist a countably k-rectifiable set $M_s \subset \mathbb{R}^n$ and a function $\theta_s : M_s \to \mathbb{N}$ such that $V_s = \theta_s |M_s|$ and $\|V_s\| = \theta_s \mathcal{H}^k \llcorner_{M_s}$, which is μ_s. Once we know μ_s is of this form, then μ_s uniquely determines V_s. So for a.e. $s \in [0, T)$, we have a unique $V_s \in \mathbf{IV}_k(U)$ such that $\|V_s\| = \mu_s$. We define $V_s \in \mathbf{V}_k(U)$ for other $s \in [0, T)$ so that $\|V_s\| = \mu_s$, for example, $V_s(\phi) = \int_U \phi(x, S_0) \, d\mu_s(x)$ for $\phi \in C_c(\mathbf{G}_k(U))$, where $S_0 \in \mathbf{G}(n, k)$ is a fixed element. Note that this is not a rectifiable varifold, but we do not care since the measure of such a set of times is zero. The resulting family $\{V_s\}_{s \in [0,T)}$ belongs to $\mathbf{IV}_k(U)$ for a.e. $s \in [0, T)$ and satisfies the claimed convergence properties.

Let a subsequence $\{i'_j\}_{j \in \mathbb{N}}$ be as above such that $V_s^{i'_j} \to V_s$ as varifolds. We may assume that this subsequence in addition achieves the same lim inf of (3.8). Let $g \in C_c^1(U; \mathbb{R}^n)$ be arbitrary. Then,

$$\delta V_s(\phi g) = \lim_{j \to \infty} \delta V_s^{i'_j}(\phi g) = -\lim_{j \to \infty} \int_U h(V_s^{i'_j}, \cdot) \cdot \phi g \, d\|V_s^{i'_j}\|$$

$$\leq \liminf_{j \to \infty} \Big(\int_U |h(V_s^{i'_j}, \cdot)|^2 \phi \, d\|V_s^{i'_j}\| \Big)^{\frac{1}{2}} \cdot \Big(\int_U |g|^2 \phi \, d\|V_s\| \Big)^{\frac{1}{2}}.$$

The above inequality shows that $\|\delta V_s\| \ll \|V_s\|$ on U since $\|\delta V_s\|(\phi) = \sup_{\|g\|_{L^\infty} \leq 1} \delta V_s(\phi g)$. Thus $h(V_s, x)$ exists and satisfies (recalling that i'_j satisfies the same lim inf as i_j)

$$\int_U |h(V_s, x)|^2 \phi \, d\|V_s\| \leq \liminf_{j \to \infty} \int_U |h(V_s^{i_j}, \cdot)|^2 \phi \, d\|V_s^{i_j}\|. \tag{3.9}$$

Since this holds for a.e. $s \in [0, T)$, we have

$$\int_{t_1}^{t_2} \int_U |h(V_s, \cdot)|^2 \phi \, d\|V_s\| \leq \liminf_{j \to \infty} \int_{t_1}^{t_2} \int_U |h(V_s^{i_j}, \cdot)|^2 \phi \, d\|V_s^{i_j}\| \, dt < \infty$$

for any $0 \leq t_1 < t_2 \leq T$, and the same inequality holds for $\phi \in C_c^1(U \times [0, T); \mathbb{R}^+)$ by a suitable approximation argument.

We only need to prove that $\{V_s\}_{s \in [0,T)}$ satisfies the integral inequality of a Brakke flow. Fix $\phi \in C_c^2(U \times [0, T); \mathbb{R}^+)$ and $0 \leq t_1 < t_2 < T$. By (3.1),

$$\int_U h(V_s^{i_j}, \cdot) \cdot \nabla\phi - \phi |h(V_s^{i_j}, \cdot)|^2 \, d\|V_s^{i_j}\| \leq \frac{1}{2} \int_{\{\phi > 0\}} \frac{|\nabla\phi|^2}{\phi} \, d\|V_s^{i_j}\| \leq C,$$

where $C = \sup \|\nabla^2 \phi\| \cdot \sup_{i \in \mathbb{N}, s \in [0,T)} \|V_s^i\|(\mathrm{spt}\,\phi)$ is bounded independently of s and j. By Fatou's Lemma,

$$\int_{t_1}^{t_2} \liminf_{j \to \infty} \Big(C + \int_U \phi |h(V_s^{i_j}, \cdot)|^2 - h(V_s^{i_j}, \cdot) \cdot \nabla\phi \, d\|V_s^{i_j}\| \Big) ds$$

$$\leq \liminf_{j \to \infty} \int_{t_1}^{t_2} \Big(C + \int_U \phi |h(V_s^{i_j}, \cdot)|^2 - h(V_s^{i_j}, \cdot) \cdot \nabla\phi \, d\|V_s^{i_j}\| \Big) ds. \tag{3.10}$$

From (3.10), we have

$$\int_{t_1}^{t_2} \liminf_{j \to \infty} \int_U \phi |h(V_s^{ij}, \cdot)|^2 - h(V_s^{ij}, \cdot) \cdot \nabla\phi \, d\|V_s^{ij}\| ds$$

$$\leq \liminf_{j \to \infty} \int_{t_1}^{t_2} \int_U \phi |h(V_s^{ij}, \cdot)|^2 - h(V_s^{ij}, \cdot) \cdot \nabla\phi \, d\|V_s^{ij}\| ds. \tag{3.11}$$

Since V_s^{ij} is a Brakke flow, the right-hand side of (3.11) is equal to

$$\liminf_{j \to \infty} \left(\|V_{t_1}^{ij}\|(\phi(\cdot, t_1)) - \|V_{t_2}^{ij}\|(\phi(\cdot, t_2)) + \int_{t_1}^{t_2} \int_U \frac{\partial\phi}{\partial t} \, d\|V_s^{ij}\| ds \right)$$

$$= \|V_{t_1}\|(\phi(\cdot, t_1)) - \|V_{t_2}\|(\phi(\cdot, t_2)) + \int_{t_1}^{t_2} \int_U \frac{\partial\phi}{\partial t} \, d\|V_s\| ds, \tag{3.12}$$

where we used the convergence of $\|V_s^{ij}\|$ to $\|V_s\|$ for all s. From (3.11) and (3.12), we will finish the proof if we prove

$$\int_U \phi |h(V_s, \cdot)|^2 - h(V_s, \cdot) \cdot \nabla\phi \, d\|V_s\|$$

$$\leq \liminf_{j \to \infty} \int_U \phi |h(V_s^{ij}, \cdot)|^2 - h(V_s^{ij}, \cdot) \cdot \nabla\phi \, d\|V_s^{ij}\| \tag{3.13}$$

for a.e. $s \in [t_1, t_2]$ (and by justifying the inequality for $\phi \in C_c^1(U \times [0, T); \mathbb{R}^+)$ which can be done by approximation). Note that the lower semicontinuity of the first term is already proved. The continuity under convergence of the second term can be harmlessly proved if V_s^{ij} converges to V_s as varifolds on U but the subsequence achieving the lim inf of (3.13) may not have this property. A technical point is the following. If we choose a subsequence $\{i'_j\}_{j\in\mathbb{N}} \subset \{i_j\}_{j\in\mathbb{N}}$ such that the lim inf is replaced by lim in (3.13), we know that $\int_U \phi |h(\cdot, V_s^{i'_j})|^2 \, d\|V_s^{i'_j}\|$ is uniformly bounded and thus we know that $V_s^{i'_j}$ converges to V_s on $\{\phi > 0\}$. But this is not enough to conclude $\int_U h(\cdot, V_s^{i'_j}) \cdot \nabla\phi \, d\|V_s^{i'_j}\|$ converges to $\int_U h(\cdot, V_s) \cdot \nabla\phi \, d\|V_s\|$ since $\nabla\phi$ does not have a compact support in $\{\phi > 0\}$. For this, we may modify the above argument slightly. Let $\psi \in C_c^2(U; \mathbb{R}^+)$ with $\mathrm{spt}\,\phi \subset \{\psi > 0\}$ be fixed. Let $\varepsilon > 0$ be arbitrary and we use $\phi_\varepsilon := \phi + \varepsilon\psi$ in the above computations. Let $\{i'_j\}_{j\in\mathbb{N}}$ be a subsequence such that

$$\liminf_{j \to \infty} \int_U \phi_\varepsilon |h(V_s^{ij}, \cdot)|^2 - h(V_s^{ij}, \cdot) \cdot \nabla\phi_\varepsilon \, d\|V_s^{ij}\|$$

$$= \lim_{j \to \infty} \int_U \phi_\varepsilon |h(V_s^{i'_j}, \cdot)|^2 - h(V_s^{i'_j}, \cdot) \cdot \nabla\phi_\varepsilon \, d\|V_s^{i'_j}\| < \infty. \tag{3.14}$$

For this subsequence, one can obtain (3.7) for ϕ_ε since we have the point-wise inequality

$$\phi_\varepsilon |h|^2 - h \cdot \nabla\phi_\varepsilon \geq \frac{1}{2}|h|^2\phi_\varepsilon - \frac{1}{2}\frac{|\nabla\phi_\varepsilon|^2}{\phi_\varepsilon}.$$

Then using Theorem 1.15 on $\{\psi > 0\}$, we conclude that $V_s^{i'_j}$ converges to V_s as varifolds on $\{\psi > 0\}$. We have

$$\int_U \phi_\varepsilon |h(V_s^{i'_j}, \cdot)|^2 - h(V_s^{i'_j}, \cdot) \cdot \nabla\phi_\varepsilon \, d\|V_s^{i'_j}\|$$

$$\geq \int_U \phi |h(V_s^{i'_j}, \cdot)|^2 - h(V_s^{i'_j}, \cdot) \cdot \nabla\phi \, d\|V_s^{i'_j}\| - \frac{\varepsilon}{2}\int_{\{\psi>0\}} \frac{|\nabla\psi|^2}{\psi} \, d\|V_s^{i'_j}\|,$$
(3.15)

where the last term is bounded uniformly by a constant times ε. Because of the varifold convergence of $V_s^{i'_j}$ to V_s on $\{\psi > 0\}$ which includes spt ϕ, we have

$$\int_U h(V_s, \cdot) \cdot \nabla\phi \, d\|V_s\| = -\int_{\mathbf{G}_k(U)} S \cdot \nabla^2\phi \, dV_s(\cdot, S)$$

$$= -\lim_{j\to\infty}\int_{\mathbf{G}_k(U)} S \cdot \nabla^2\phi \, dV_s^{i'_j}(\cdot, S) = \lim_{j\to\infty}\int_U h(V_s^{i'_j}, \cdot) \cdot \nabla\phi \, d\|V_s^{i'_j}\|.$$
(3.16)

Now (3.14), (3.15) and (3.16) as well as (3.9) show

$$\liminf_{j\to\infty}\int_U \phi_\varepsilon |h(V_s^{i_j}, \cdot)|^2 - h(V_s^{i_j}, \cdot) \cdot \nabla\phi_\varepsilon \, d\|V_s^{i_j}\|$$

$$\geq \int_U \phi |h(V_s, \cdot)|^2 - h(V_s, \cdot) \cdot \nabla\phi \, d\|V_s\| - c(\|\psi\|_{C^2}, \text{spt}\,\psi)\varepsilon,$$
(3.17)

which holds for a.e. $s \in [0, T]$. Finally, with (3.11) and (3.12) (where ϕ there replaced by ϕ_ε) as well as (3.17), we obtain

$$\|V_{t_1}\|(\phi_\varepsilon(\cdot, t_1)) - \|V_{t_2}\|(\phi_\varepsilon(\cdot, t_2)) + \int_{t_1}^{t_2}\int_U \frac{\partial\phi_\varepsilon}{\partial t} \, d\|V_s\| \, ds$$

$$\geq \int_{t_1}^{t_2}\int_U \phi |h(V_s, \cdot)|^2 - h(V_s, \cdot) \cdot \nabla\phi \, d\|V_s\| \, ds - c(\|\psi\|_{C^2}, \text{spt}\,\psi)(t_2 - t_1)\varepsilon$$

and by letting $\varepsilon \to 0$, we obtain the desired inequality for $\phi \in C_c^2(U \times [0, T]; \mathbb{R}^+)$, and for any $\phi \in C_c^1(U \times [0, T]; \mathbb{R}^+)$ by approximation. $\square$

The compactness property of a Brakke flow is not stated in [7], but a result in the same spirit can be found in [7, Chapter 4]. The precise compactness statement can be found in Ilmanen's paper [19].

3.4   Tangent Flows

Given a k-dimensional Brakke flow $\{V_t\}_{t \in [0,T)}$ in $U \subset \mathbb{R}^n$, we have a notion of tangent flow at any given point in space–time which is defined as follows. Let $\{r_i\}_{i \in \mathbb{N}}$ be an arbitrary sequence of positive numbers converging to 0. Let $x_0 \in U$ and $t_0 \in (0, T)$ be fixed and assume without loss of generality that $x_0 = 0$ and $t_0 = 0$ after change of variables. We consider a parabolic rescaling of V_t centered at $(0, 0)$. This means that we rescale x by $1/r_i$ and t by $1/r_i^2$. If the Brakke flow is $V_t = |M_t|$, then the rescaled flow would be

$$M_s^i := r_i^{-1} M_{r_i^2 s}.$$

This M_s^i is a set in $U^i := \{z \in \mathbb{R}^n : z = r_i^{-1} x, \ x \in U\}$. In the language of varifolds, the corresponding varifold is defined for $\phi \in C_c(\mathbf{G}_k(U^i))$:

$$V_s^i(\phi) : = |M_s^i|(\phi) = \int_{M_s^i} \phi(z, \mathrm{Tan}_z M_s^i) \mathrm{d}\mathcal{H}^k(z)$$

$$= \frac{1}{r_i^k} \int_{M_t} \phi(r_i^{-1} x, \mathrm{Tan}_x M_t) \, \mathrm{d}\mathcal{H}^k(x),$$

where we changed the variables as $t = r_i^2 s$ and $x = r_i z$. If we define

$$\tau_{r_i}(x) := r_i^{-1} x,$$

using the notation of push-forward of varifolds, we see that $V_s^i = (\tau_{r_i})_\sharp V_{r_i^2 s}$. Likewise for a general Brakke flow, let

$$V_s^i(\phi) := \frac{1}{r_i^k} \int_U \phi(r_i^{-1} x, S) \, \mathrm{d}V_{r_i^2 s}(x, S) = (\tau_{r_i})_\sharp V_{r_i^2 s}(\phi)$$

for $\phi \in C_c(\mathbf{G}_k(U^i))$. Then the rescaled varifolds again give Brakke flows which can be checked directly (note that the generalized mean curvature has $h(V_s^i, z) = r_i h(V_t, r_i z)$). For such a sequence, we have the following:

Theorem 3.8 ([20]) *Suppose that $\{V_t\}_{t \in [0,T)}$ is a Brakke flow on U, $x_0 \in U$, $t_0 \in (0, T)$ and $\{r_i\}_{i \in \mathbb{N}}$ are given. Suppose $\{V_t^i\}$ is defined as above. Then there exist a subsequence $\{i_j\}_{j \in \mathbb{N}}$ and a Brakke flow $\{\tilde{V}_t\}_{t \in \mathbb{R}}$ such that $\{V_t^{i_j}\}_{j \in \mathbb{N}}$ converge to $\{\tilde{V}_t\}$*

locally on $\mathbb{R}^n \times \mathbb{R}$ as described in Theorem 3.7. Moreover, for any $r > 0$ and writing $\tau_r(x) := r^{-1}x$, we have

$$\|(\tau_r)_\sharp \tilde{V}_{-r^2}\| = \|\tilde{V}_{-1}\|$$

for all $r > 0$ and

$$h(\tilde{V}_s, x) = \frac{S^\perp(x)}{2s} \tag{3.18}$$

for $\tilde{V}_s$ a.e. on $\mathbf{G}_k(\mathbb{R}^n)$ for a.e. $s < 0$.

Proof We may assume that $x_0 = 0$ and $t_0 = 0$. Suppose that $B_{3R} \times [-R^2, R^2]$ is included in the domain of the Brakke flow. For any fixed $r > 0$, we have for all sufficiently large i and $s \in [-r^2, r^2]$

$$\|V_s^i\|(B_r) = r_i^{-k}\|V_{r_i^2 s}\|(B_{r_i r}) \le c(R, k)r^k \sup_{t \in [-R^2, R^2]} \|V_t\|(B_R) =: \hat{c}r^k \tag{3.19}$$

by Proposition 3.5. Thus, by Theorem 3.7, we have a converging subsequence (denoted by the same index) and a limit Brakke flow $\{\tilde{V}_t\}_{t \in \mathbb{R}}$ on $\mathbb{R}^n$. Because of (3.19), we also have

$$\sup_{r > 0, s \in \mathbb{R}} r^{-k}\|\tilde{V}_s\|(B_r) \le \hat{c}. \tag{3.20}$$

Next take the formula (3.6) with $y = 0$ and $s = 0$, where $\hat{\rho} = \rho\zeta$ is just as what was defined there. Since the last term of (3.6) is bounded by $R^{-2}(t_2 - t_1)\exp(R^2/t_1)$ times constant for all small $t_1 < t_2 < 0$,

$$a := \lim_{t \to 0-} \int_{B_{3R}} \hat{\rho}_{(0,0)}(x, t)\, d\|V_t\|$$

exists. This means that for any $s < 0$, we have

$$\int_{B_{3R}} \hat{\rho}_{(0,0)}(x, r_i^2 s)\, d\|V_{r_i^2 s}\|(x) = \int_{B_{3R/r_i}} \rho_{(0,0)}(z, s)\zeta(r_i z)\, d\|V_s^i\|(z) \to a \tag{3.21}$$

as $i \to \infty$. We now claim that for all $s < 0$, we have

$$\int_{\mathbb{R}^n} \rho_{(0,0)}(z, s)\, d\|\tilde{V}_s\|(z) = a. \tag{3.22}$$

This is expected from (3.21) since $\|V_s^i\|$ converges to $\|\tilde{V}_s\|$ and the growth of the measures is polynomial, while $\rho_{(0,0)}(z, s)$ decays exponentially near infinity. We write out the details of the estimate for the readers' convenience in the next few lines. For any large and fixed $\hat{r} > 0$, for all sufficiently large i so that $R/r_i > \hat{r}$, we have

$$\left| \int_{B_{\hat{r}}} \rho_{(0,0)}(z,s) \, d\| V_s^i \| - \int_{B_{3R/r_i}} \rho_{(0,0)}(z,s) \zeta(r_i z) \, d\| V_s^i \|(z) \right|$$

$$\leq \int_{B_{3R/r_i} \setminus B_{\hat{r}}} \rho_{(0,0)}(z,s) \, d\| V_s^i \|(z)$$

$$= \int_0^\infty \| V_s^i \| ((B_{3R/r_i} \setminus B_{\hat{r}}) \cap \{ z : \rho_{(0,0)}(z,s) > l \}) \, dl.$$

Here we used $\int f \, d\mu = \int_0^\infty \mu(\{ f > l \}) \, dl$ for general $f \geq 0$, which follows from the Fubini theorem. The explicit computations using (3.19) and a change of variable show that

$$\leq \hat{c} \left(\frac{3R}{r_i} \right)^k \frac{\exp \frac{(3R)^2}{4sr_i^2}}{(4\pi(-s))^{\frac{k}{2}}} + \frac{\hat{c}}{-2s} \int_{\hat{r}}^{3R/r_i} \frac{r^{k+1} \exp(\frac{r^2}{4s})}{(4\pi(-s))^{\frac{k}{2}}} \, dr,$$

which is uniformly small for large $\hat{r}$ and on $0 > s \geq -L$ for any fixed $L > 0$. This estimate proves that, for given any $\varepsilon > 0$, there exists $\hat{r}$ such that

$$\left| \int_{B_{\hat{r}}} \rho_{(0,0)}(z,s) \, d\| V_s^i \| - \int_{B_{3R/r_i}} \rho_{(0,0)}(z,s) \zeta(r_i z) \, d\| V_s^i \|(z) \right| < \varepsilon \qquad (3.23)$$

for all sufficiently large i. The same type of truncation at infinity holds for $\| \tilde{V}_s \|$ as well. These estimates and (3.21) implies (3.22). Since $\tilde{V}_t$ is a Brakke flow, for any fixed $t_1 < t_2 < 0$,

$$\int_{\mathbb{R}^n} \rho_{(0,0)}(z,s) \, d\| \tilde{V}_s \| \Big|_{s=t_1}^{t_2}$$

$$\leq - \int_{t_1}^{t_2} \int_{\mathbf{G}_k(\mathbb{R}^n)} \left| h(\tilde{V}_s, z) - \frac{S^\perp(z)}{2s} \right|^2 \rho_{(0,0)}(z,s) \, d\tilde{V}_s ds. \qquad (3.24)$$

In fact, $\rho_{(0,0)}$ does not have a compact support, but we can easily justify the above by approximation. Since the left-hand side is 0 by (3.22), we can conclude that (3.18) holds. To prove the invariance, define for $s < 0$

$$W_s := (\tau_{\sqrt{-s}})_\sharp \tilde{V}_s.$$

Recall that this means that, if $\tilde{V}_s = |M_s|$, we would be looking at $\frac{1}{\sqrt{-s}} M_s$. If we prove that $\| W_s \|(\phi)$ is constant for all $\phi \in C_c^1(\mathbb{R}^n; \mathbb{R}^+)$ in s, we finish the proof. For $0 > t_2 > t_1$ and $\phi \in C_c^1(\mathbb{R}^n; \mathbb{R}^+)$, we have

$$\|W_{t_2}\|(\phi) - \|W_{t_1}\|(\phi)$$

$$= (-t_2)^{-\frac{k}{2}} \int \phi\left(\frac{x}{\sqrt{-t_2}}\right) d\|\tilde{V}_{t_2}\| - (-t_1)^{-\frac{k}{2}} \int \phi\left(\frac{x}{\sqrt{-t_1}}\right) d\|\tilde{V}_{t_1}\|. \tag{3.25}$$

Regarding $(-t)^{-\frac{k}{2}}\phi(x/\sqrt{-t})$ as a test function in (2.11), we obtain

$$\leq \int_{t_1}^{t_2} \int_{\mathbf{G}_k(\mathbb{R}^n)} (-t)^{-\frac{k}{2}}\left(\frac{k\phi}{2(-t)} - \phi|h|^2 + \frac{\nabla\phi \cdot h}{\sqrt{-t}} - \frac{\nabla\phi \cdot x}{2t\sqrt{-t}}\right) d\tilde{V}_t \, dt. \tag{3.26}$$

We use (3.18) and Theorem 1.18 to obtain for the second term of (3.26)

$$\int \phi|h|^2 \, d\tilde{V}_t = \int \frac{\phi h \cdot S^\perp(x)}{2t} \, d\tilde{V}_t = \int \frac{\phi h \cdot x}{2t} \, d\tilde{V}_t$$

$$= -\int \frac{S \cdot \nabla(x\phi)}{2t} \, d\tilde{V}_t. \tag{3.27}$$

Since $S \cdot \nabla(x\phi) = \phi S \cdot I + S \cdot (\nabla\phi \otimes x)/\sqrt{-t} = k\phi + \nabla\phi \cdot S(x)/\sqrt{-t}$, we obtain from the above that

$$\|W_{t_2}\|(\phi) - \|W_{t_1}\|(\phi)$$

$$\leq \int_{t_1}^{t_2} \int (-t)^{-\frac{k}{2}}\left(\frac{\nabla\phi \cdot S(x)}{2t\sqrt{-t}} + \frac{\nabla\phi \cdot h}{\sqrt{-t}} - \frac{\nabla\phi \cdot x}{2t\sqrt{-t}}\right) d\tilde{V}_t dt. \tag{3.28}$$

Again using (3.18) and noticing that $S(x) + S^\perp(x) = x$, we see that the right-hand side of (3.28) is 0, which shows that $\|W_t\|(\phi)$ is non-increasing in t. To prove the opposite inequality, observe first that we have

$$\int \rho_{(0,0)}(x, t) \, d\|\tilde{V}_t\| = (-t)^{-\frac{k}{2}} \int \rho_{(0,0)}\left(\frac{x}{\sqrt{-t}}, -1\right) d\|\tilde{V}_t\|$$

$$= \int \rho_{(0,0)}(x, -1) \, d\|W_t\|,$$

which is independent of t due to (3.22). Let $0 > t_2 > t_1$ and $\phi \in C_c^1(\mathbb{R}^n; \mathbb{R}^+)$ be arbitrary again. Choose a small $c > 0$ so that $\rho_{(0,0)}(x, -1) \geq c\phi(x)$ for all $x \in \mathbb{R}^n$. We then use $\rho_{(0,0)}(x, -1) - c\phi(x)$ in place of ϕ in (3.25). We proceed by the same computations as before. This test function does not have a compact support, but we can justify the use in (2.11) by the exponential decay of $\rho_{(0,0)}$ and suitable approximation. The computation of the right-hand side for $\rho_{(0,0)}$ is just like the one for Huisken's monotonicity formula, and the one for ϕ also vanishes. This shows that $\|W_t\|(\rho_{(0,0)} - c\phi)$ is non-increasing in t. Since $\|W_t\|(\rho_{(0,0)})$ is independent of t, this shows that $\|W_t\|(\phi)$ is non-decreasing in t. Combined with the first part, this shows that $\|W_t\|(\phi)$ is independent of t and shows the invariance of $\|W_t\|$ in t as a measure. $\qquad\square$

In the case that $V \in \mathbf{IV}_k(U)$ is stationary in U, i.e., $\delta V = 0$, then V is a time-independent Brakke flow and above $\tilde{V}$ is called a *tangent cone*. In this case, since we may simply drop the time-dependence in Theorem 3.8 and $h(\tilde{V}, x) = 0$, $\tilde{V}$ satisfies $\|(\tau_r)_\sharp \tilde{V}\| = \|\tilde{V}\|$ for all $r > 0$ and $S^\perp(x) = 0$ for $\tilde{V}$ a.e. on $\mathbf{G}_k(\mathbb{R}^n)$. While the dilation invariance holds for this tangent cone, the parabolic dilation invariance holds only backward in time for the tangent flow.

If a Brakke flow happened to be a classical MCF in a space–time neighborhood of $(x_0, t_0) \in U \times (0, T)$, it is clear that the tangent flow is a time-independent k-dimensional subspace with single multiplicity, that is, $\tilde{V}_s = |S|$ for some $S \in \mathbf{G}(n, k)$. It is a nice corollary of the regularity theorem of Chap. 6 that the converse is also true, that is, if a Brakke flow V_t has a time-independent tangent flow $\tilde{V}_s = |S|$ at (x_0, t_0), then $\mathrm{spt} \, \|V_t\|$ is a smooth classical MCF in a space–time neighborhood of (x_0, t_0). Because of this property, the set of regular points of a Brakke flow can be characterized by the existence of such a tangent flow. The singular set of a Brakke flow is the complement of the regular points where other kinds of tangent flow appear. There are a number of properties of tangent flows which follow from Huisken's monotonicity formula, and I refer the reader to [39] for further reading. I just mention that one can stratify the singular points depending on the dimension of the homogeneous subspace of tangent flows. Their Hausdorff dimensions can be estimated in terms of such dimensions using Federer's dimension–reducing argument. These aspects for area-minimizing currents have been successfully studied, where one has area-minimizing cones. For more refined information on the structure of singular sets, see more recent work [8].

Chapter 4
A General Existence Theorem for a Brakke Flow in Codimension One

4.1 Main Existence Result

One of the cornerstone results in [7] is the general existence theorem of a Brakke flow. For any $1 \leq k < n$ and any initial rectifiable k-varifold with some minor assumption, Brakke gave a proof of a time-global existence of rectifiable Brakke flow starting from the given data. When the initial data is an integral k-varifold, the obtained flow is also integral in the sense defined in Chap. 2. Even after close to 40 years, it still stands as a remarkable insight, with an ingenious approximation scheme and fine compactness-type theorems of varifolds. On the other hand, one serious concern on his existence result is that there is no guarantee that the Brakke flow he obtained is non-trivial when the initial data is not a smooth k-dimensional surface. That is, even though one can see that his approximation scheme produces a reasonable family of approximate Brakke flows, it is not clear if the Brakke flow obtained as the limit of his approximation is non-zero for $t > 0$. An easy fact from the definition of Brakke flow is that $V_t = 0$ for $t > 0$ is always a Brakke flow for any given initial data at $t = 0$. It also has to be said that following Brakke's proof in detail poses a serious hurdle for even the most dedicated and interested specialists in geometric measure theory. This issue was left untouched for several years. It inevitably appears that there has not been any research result based directly on his existence theorem in the past, despite the fundamental nature of the question.

With these points in mind, in this chapter we present our general existence results of the Brakke flow in the case of codimension one which always gives a non-trivial flow [22]. The work is a careful reworking of [7, Chapter 4] with several new inputs which solve this issue. If we state the claim in a simple manner, the following is the essential content.

© The Author(s), under exclusive license to Springer Nature Singapore Pte Ltd. 2019   49
Y. Tonegawa, *Brakke's Mean Curvature Flow*, SpringerBriefs in Mathematics,
https://doi.org/10.1007/978-981-13-7075-5_4

Theorem 4.1 *Suppose that $\Gamma_0 \subset \mathbb{R}^{n+1}$ is a closed countably n-rectifiable set such that $\mathbb{R}^{n+1} \setminus \Gamma_0$ is not connected. Assume that $\int_{\Gamma_0} \exp(-c|x|) \, d\mathcal{H}^n(x) < \infty$ for some $c \geq 0$. Then there exists a non-trivial Brakke flow $\{V_t\}_{t \geq 0}$ starting from Γ_0, i.e. $V_0 = |\Gamma_0|$.*

General closed countably n-rectifiable sets can be quite singular with cusps and corners. For $n = 1$, Γ_0 can be a network-like set, and for $n = 2$, Γ_0 can be a soap-bubble cluster. Also Γ_0 can have infinite $\mathcal{H}^n$ measure as long as it grows at most exponentially near infinity, and we require that there are more than one connected component for the complement.

The most interesting aspect is that the result gives general and time-global existence results which allow triple junction singularities and their topological changes in a natural way. This is a stark difference from the level set method [9, 13], where MCF is represented as a boundary of one domain so that the presence of triple junctions cannot be handled in a simple manner. If Γ_0 is not smooth, the solution obtained by the level set method may develop the so-called fattening immediately, while our existence results give a Brakke flow which by definition has locally finite $\mathcal{H}^n$ measure. On the other hand, the solution may not be unique in general with the presence of singularities, and one can see non-uniqueness with a simple example.

The basic idea of how to prevent the flow from being trivial is simple. Instead of looking at the varifolds only as Brakke did, we look at moving domains whose boundaries move by the mean curvature. If we know that these domains move continuously in some sense, we can be sure that there will be non-empty boundaries. This also suits well the physical picture of grain boundary motion, where each grain is represented as one of these moving domains. In more precise language, given Γ_0 as stated above, we first choose a set of disjoint non-empty open sets

$$E_{0,1}, \ldots, E_{0,N} \subset \mathbb{R}^{n+1}$$

for some $N \geq 2$ such that

$$\mathbb{R}^{n+1} \setminus \Gamma_0 = \cup_{i=1}^{N} E_{0,i}.$$

Then, with this choice, along with the existence of a Brakke flow $\{V_t\}_{t \geq 0}$ with

$$V_0 = |\Gamma_0|,$$

we show that there exists a family of open sets $\{E_i(t)\}_{t \geq 0}$ for each $i = 1, \ldots, N$ with the following properties. We have

$$E_i(0) = E_{0,i} \text{ for } i = 1, \ldots, N,$$

$$E_1(t), \ldots, E_N(t) \text{ are disjoint for each } t \geq 0,$$

and setting

$$\Gamma(t) := \mathbb{R}^{n+1} \setminus \cup_{i=1}^{N} E_i(t)$$

and defining a space–time measure

$$d\mu := d\|V_t\| dt \quad \text{in } \mathbb{R}^{n+1} \times [0, \infty),$$

we have

$$\{x \in \mathbb{R}^{n+1} : (x, t) \in \operatorname{spt} \mu\} = \Gamma(t) \text{ for all } t > 0.$$

Here, $\operatorname{spt} \mu$ is like a history of motion of the flow and the time-slice coincides with the union of the boundaries of $E_i(t)$. Moreover, $E_i(t)$ moves continuously in time with respect to the Lebesgue measure. That is, for any ball $B_r(x)$ in $\mathbb{R}^{n+1}$,

$$\mathcal{L}^{n+1}(B_r(x) \cap E_i(t))$$

is a $C^0([0, \infty)) \cap C_{loc}^{\frac{1}{2}}(0, \infty)$ function in time. As we stated above, the picture is that each $E_i(t)$ is a "grain" which grows or shrinks in time, and the union of their boundaries $\Gamma(t)$ moves by mean curvature in the sense of Brakke, if loosely speaking. Note that the choice of $E_{0,1}, \ldots, E_{0,N}$ is not unique. It is like assigning a number to each component of $\mathbb{R}^{n+1} \setminus \Gamma_0$, and we may (and we should) end up with a different flow depending on the assignment. We only ask that $N \geq 2$, since $N = 1$ ($E_{0,1} = \mathbb{R}^{n+1} \setminus \Gamma_0$) will produce a trivial Brakke flow $V_t = 0$ for all $t > 0$ in our scheme. Since $\mathbb{R}^{n+1} \setminus \Gamma_0$ is assumed to be non-connected, we may always choose a set of initial open sets so that $N \geq 2$. The continuous family $E_i(t)$ ensures a certain sense of continuity of V_t, in the sense that V_t cannot vanish arbitrarily. In particular, unless all but one $E_i(t)$ become empty sets, we have $V_t \neq 0$.

In the following, we give an outline of the proof and we restrict ourselves to the case of $\mathcal{H}^n(\Gamma_0) < \infty$, since some of the presentation become simpler without losing much of the essential feature. It is meant to give an intuitive picture of the proof for interested readers who may wish to understand the content of [22].

4.2 A First Try, and Its Problems

In this section, I will describe a naive but reasonable method to construct an approximate mean curvature flow starting from a given closed countably n-rectifiable set Γ_0. This is going to be a failed attempt but it explains why we want to do what we will do next.

As explained before, given such a Γ_0, we choose a set of disjoint open sets $E_{0,1}, \ldots, E_{0,N}, N \geq 2$, such that $\mathbb{R}^{n+1} \setminus \Gamma_0 = \cup_{i=1}^{N} E_{0,i}$. Since $\mathcal{H}^n(\Gamma_0) < \infty$, Γ_0

cannot have any interior point, so that $\Gamma_0 = \cup_{i=1}^N \partial E_{0,i}$ in particular. Fix a time step size to be Δt, and we construct a time-discrete approximate mean curvature flow starting from Γ_0. The hope is that we obtain a Brakke flow as a limit of this sequence of approximate mean curvature flows as $\Delta t \to 0$. Since Γ_0 is only assumed to be closed and countably n-rectifiable, Γ_0 cannot have the usual mean curvature vector, or even worse, $|\Gamma_0|$ may not even have a bounded first variation. Thus, we define an analogue of the mean curvature vector good for a countably n-rectifiable set. Suppose for a moment that Γ_0 is smooth and without boundary. We use (1.13) to motivate the approximate mean curvature vector. For that purpose, let $\varepsilon > 0$ be given and define

$$\hat{\Phi}_\varepsilon(x) = \frac{1}{(2\pi\varepsilon^2)^{\frac{n+1}{2}}} \exp\left(-\frac{|x|^2}{2\varepsilon^2}\right), \qquad \Phi_\varepsilon(x) = c(\varepsilon)\psi(x)\hat{\Phi}_\varepsilon(x), \qquad (4.1)$$

where ψ is a radially symmetric function such that $\psi(x) = 1$ for $|x| \leq 1/2$, $\psi(x) = 0$ for $|x| \geq 1$, $0 \leq \psi(x) \leq 1$, $|\nabla\psi(x)| \leq 3$ and $|\nabla^2\psi(x)| \leq 9$ for all $x \in B_1$. $c(\varepsilon)$ is chosen so that $\int_{\mathbb{R}^{n+1}} \Phi_\varepsilon(x)\,\mathrm{d}x = 1$. The function Φ_ε is a truncated Gaussian function[1] which converges to the delta function as $\varepsilon \to 0$. In (1.13) and for $l = 1, \ldots, n+1$, we use $g(x) = \Phi_\varepsilon(x-y)e_l$, where e_l is the standard basis with its l-th component being equal to 1. Then we have

$$\int_{\Gamma_0} \sum_{j=1}^{n+1} (T_x\Gamma_0)_{lj} \nabla_{x_j} \Phi_\varepsilon(x-y)\,\mathrm{d}\mathcal{H}^k(x) = -\int_{\Gamma_0} h(\Gamma_0,x)_l \Phi_\varepsilon(x-y)\,\mathrm{d}\mathcal{H}^k(x)$$

$$(4.2)$$

where $h(\Gamma_0,x)_l$ is the l-th component of $h(\Gamma_0,x)$ and the integrand on the left-hand side of (4.2) is the l-th component of $(T_x\Gamma_0) \circ \nabla\Phi_\varepsilon$. For any $y \in \Gamma_0$, if we divide both sides of (4.2) by $\int_{\Gamma_0} \Phi_\varepsilon(x-y)\,\mathrm{d}\mathcal{H}^k(x)$ and if we let $\varepsilon \to 0$, then the right-hand side should converge to $-h(\Gamma_0,y)_l$, since $\Phi_\varepsilon(x-y)$ will be concentrated around y. Note that the left-hand side of (4.2) is well-defined even for a countably n-rectifiable Γ_0. Thus it is reasonable to define

$$\tilde{h}_\varepsilon(\Gamma_0,y) := -\frac{\int_{\Gamma_0} (T_x\Gamma_0) \circ \nabla\Phi_\varepsilon(x-y)\,\mathrm{d}\mathcal{H}^k(x)}{\int_{\Gamma_0} \Phi_\varepsilon(x-y)\,\mathrm{d}\mathcal{H}^k(x) + \varepsilon} \qquad (4.3)$$

as an approximate mean curvature vector of Γ_0 at y. We add ε in the denominator so that $\tilde{h}_\varepsilon$ behaves nicely away from Γ_0. Note that $\int_{\Gamma_0} \Phi_\varepsilon(x-y)\,\mathrm{d}\mathcal{H}^k(x) \to 0$ if $y \notin \Gamma_0$, and having ε in the denominator ensures that $\tilde{h}_\varepsilon(\Gamma_0,y) \to 0$ if $y \notin \Gamma_0$. It turned out that it is desirable to take another convolution (see Lemma 4.12) and we define

---

[1]This particular choice of mollifier turned out to be important in the error estimates, see Lemma 4.17.

$$h_\varepsilon(\Gamma_0, x) := \int_{\mathbb{R}^{n+1}} \Phi_\varepsilon(x - y) \tilde{h}_\varepsilon(\Gamma_0, y) \, dy. \tag{4.4}$$

Both $\tilde{h}_\varepsilon$ and h_ε are smooth vector fields on $\mathbb{R}^{n+1}$ and note that, for a smooth Γ_0, we have $h_\varepsilon(\Gamma_0, x) \to h(\Gamma_0, x)$ for $x \in \Gamma_0$ and otherwise $h_\varepsilon(\Gamma_0, x) \to 0$. For non-smooth Γ_0, they typically behave badly as $\varepsilon \to 0$. Still, it is reasonable to move Γ_0 and also $E_{0,1}, \dots, E_{0,N}$ by this approximate mean curvature vector for a short time interval of Δt. Thus define

$$F(x) = x + h_\varepsilon(\bar{\Gamma}_0, x) \Delta t$$

and define $\Gamma(0) = \Gamma_0$, $\Gamma(\Delta t) = F(\Gamma_0)$, $E_i(\Delta t) = F(E_{0,i})$ for each $i = 1, \dots, N$. As for the size of Δt, we need to choose an appropriately small Δt. Since $|\nabla^l \Phi_\varepsilon| \leq O(\varepsilon^{-2l}) \Phi_\varepsilon$ plus some exponentially small errors coming from the derivatives of ψ, one can prove the following. We omit the proof (see [22, Lemma 5.1]).

Lemma 4.2 *Suppose $\mathcal{H}^n(\Gamma_0) \leq M$. There exists a constant c depending only on n and M such that*

$$\sup_{x \in \mathbb{R}^{n+1}, \varepsilon \in (0,1)} \{\varepsilon^2 |h_\varepsilon(\Gamma_0, x)|, \varepsilon^4 |\nabla h_\varepsilon(\Gamma_0, x)|\} \leq c. \tag{4.5}$$

Thus, as long as $\Delta t \ll \varepsilon^4$, $\|F(x) - x\|_{C^1}$ is small. The map F is a diffeomorphism, and the newly defined $E_1(\Delta t), \dots, E_N(\Delta t)$ will be mutually disjoint and we have $\Gamma(\Delta t) = \cup_{i=1}^N \partial E_i(\Delta t)$. The regularity of $\Gamma(\Delta t)$ will not improve, but it stays closed and countably n-rectifiable with a finite $\mathcal{H}^n$ measure. We may repeat the same procedure for $\Gamma(\Delta t), E_1(\Delta t), \dots, E_N(\Delta t)$ by computing the approximate mean curvature vector $h_\varepsilon(\Gamma(\Delta t), x)$ and moving these sets by $F(x) = x + h_\varepsilon(\Gamma(\Delta t), x) \Delta t$. Let $\Gamma(k \Delta t), E_1(k \Delta t), \dots, E_N(k \Delta t)$ be sets defined under this procedure inductively. To continue this way, we need to make sure that (4.5) holds so that we do not worry about F being a diffeomorphism. But the constant c in (4.5) depends only on the upper bound M of $\mathcal{H}^n(\Gamma_0)$. Since we are approximating the mean curvature flow, intuitively, we should have a decreasing $\mathcal{H}^n$ measure in each step with some small error depending on ε. Or, if it may not be decreasing, we only need an upper bound on $\mathcal{H}^n(\Gamma(k \Delta t))$. This can be proved in fact with some rather careful estimates on h_ε ([22, (5.56) with $\Omega = 1$ and $c_1 = 0$]) but for now let us ignore this and continue. The procedure can be repeated until the accumulated error for $\mathcal{H}^n(\Gamma(k \Delta t))$ becomes too large and one cannot be sure if F is a diffeomorphism. We can construct such approximate flow for each $\varepsilon > 0$ and we may obtain a Brakke flow as a limit $\varepsilon, \Delta t \to 0$ while keeping $\Delta t \ll \varepsilon$. This seems like a good try.

There are two serious problems for such a procedure. The first is that the resulting domains and the boundaries in later time are all diffeomorphic to the initial ones. We would like to see some domain vanishing at some point for one thing. For the $n = 1$ case, there should be occasional collisions of triple junctions which result in junctions of more than three edges. They should then immediately split to triple junctions connected by short curves. It is unlikely that such a phenomenon can be approximated by a time-discrete sequence of diffeomorphisms. We definitely need some non-diffeomorphic maps near the singularities. The second point is that we do not have any control of what is happening on the length scale smaller than ε since the approximate mean curvature is too diffused to see anything smaller. It turns out that, with the above scheme, we have some control of the L^2 space–time integral of $\tilde{h}_\varepsilon(\Gamma(k\Delta t), x)$ with respect to the (again diffused) surface measure of $\Gamma(k\Delta t)$, thus there is a hope of having some geometric control for a length scale larger than ε. To say anything about the limit varifolds as ε, $\Delta t \to 0$, this is unfortunately not enough.

To solve these problems, it is a great idea due to Brakke that we insert a certain Lipschitz map in each discrete time step which almost and locally minimizes surface measure *only around singularities*. The Lipschitz map is designed carefully so that it is more or less the identity map away from singularities. The next section describes this map.

4.3　Open Partitions and Admissible Lipschitz Maps

In constructing a time-discrete flow, these moving domains should belong to the same class as the initial data. We thus define the following for clarification.

Definition 4.3 Fix $N \geq 2$. A finite and ordered collection of sets $\mathcal{E} = \{E_i\}_{i=1}^N$ in $\mathbb{R}^{n+1}$ is called a finite open partition of N elements if:

(a) $E_1, \ldots, E_N$ are open and disjoint,
(b) $\mathcal{H}^n(\mathbb{R}^{n+1} \setminus \cup_{i=1}^N E_i) < \infty$,
(c) $\cup_{i=1}^N \partial E_i$ is countably n-rectifiable.

The set of all finite open partition of N elements is denoted by $\mathcal{OP}^N$.

Since $\mathbb{R}^{n+1} \setminus \cup_{i=1}^N E_i$ is closed and cannot have non-trivial interior by (b), we have

$$\mathbb{R}^{n+1} \setminus \cup_{i=1}^N E_i = \cup_{i=1}^N \partial E_i.$$

One may think that $\mathbb{R}^{n+1}$ is partitioned into $E_1, \ldots, E_N$ by their boundaries $\cup_{i=1}^N \partial E_i$ which has a finite $\mathcal{H}^n$ measure. We allow that some of E_i may be empty set. Given $\mathcal{E} = \{E_i\}_{i=1}^N \in \mathcal{OP}^N$, we define

$$\partial\mathcal{E} := |\cup_{i=1}^N \partial E_i| \in \mathbf{IV}_n(\mathbb{R}^{n+1}).$$

It is a naturally defined unit density varifold from $\cup_{i=1}^N \partial E_i$ which is assumed to be countably n-rectifiable by (c). With a slight abuse of notation, we may regard $\partial\mathcal{E}$ as a unit density varifold as defined above, but also we sometimes take $\partial\mathcal{E}$ as a closed set $\cup_{i=1}^N \partial E_i$ in the following, whenever this should not cause any confusion.

We next define a class of Lipschitz maps with desirable properties. There are a few requirements for this map, given an open partition $\mathcal{E}$. We want this Lipschitz map to send an element in $\mathcal{OP}^N$ into itself. It should also permit the kind of topological changes we would like to see.

Definition 4.4 Given $\mathcal{E} = \{E_i\}_{i=1}^N \in \mathcal{OP}^N$, a function $f : \mathbb{R}^{n+1} \to \mathbb{R}^{n+1}$ is called $\mathcal{E}$-admissible if it is Lipschitz and satisfies the following. Define $\tilde{E}_i := \text{int}\,(f(E_i))$ for each i. Then:

(a) $\{\tilde{E}_i\}_{i=1}^N$ are mutually disjoint,
(b) $\mathbb{R}^{n+1} \setminus \cup_{i=1}^N \tilde{E}_i \subset f(\cup_{i=1}^N \partial E_i)$.

With this definition, we have:

Lemma 4.5 $\{\tilde{E}_i\}_{i=1}^N \in \mathcal{OP}^N$.

It is easy to check this. Each $\tilde{E}_i$ is open by definition ($\tilde{E}_i$ is a set of interior points) and $\{\tilde{E}_i\}_{i=1}^N$ are mutually disjoint by (a). By (b),

$$\mathcal{H}^n(\mathbb{R}^{n+1} \setminus \cup_{i=1}^N \tilde{E}_i) \leq \mathcal{H}^n(f(\cup_{i=1}^N \partial E_i)) \leq (\text{Lip } f)^n \mathcal{H}^n(\cup_{i=1}^N \partial E_i) < \infty,$$

where we used the property of the Hausdorff measure under the Lipschitz function. The countable rectifiability of $\mathbb{R}^{n+1} \setminus \cup_{i=1}^N \tilde{E}_i$ follows from (b) and Proposition 1.2. This completes the proof of the lemma.

Definition 4.6 For $\mathcal{E} \in \mathcal{OP}^N$ and $\mathcal{E}$-admissible function f, let $\tilde{E}_i$ be as in Definition 4.4 and define $f_\star : \mathcal{OP}^N \to \mathcal{OP}^N$ as $f_\star \mathcal{E} := \{\tilde{E}_i\}_{i=1}^N$.

Any smooth diffeomorphism with a bounded Lipschitz constant is $\mathcal{E}$-admissible, since $\tilde{E}_i = f(E_i)$ and $\mathbb{R}^{n+1} \setminus \cup_{i=1}^N \tilde{E}_i = f(\cup_{i=1}^N \partial E_i)$ in this case. Hence, for any $\mathcal{E} \in \mathcal{OP}^N$, as long as we map $\mathcal{E}$ by a diffeomorphism or a function in the $\mathcal{E}$-admissible class, it stays in $\mathcal{OP}^N$.

The choice of definition we made for the admissible class may not be unique. This admissible class is a new concept introduced in [22]. On the other hand, there seem to be few choices of such admissible class, either. We would like to have some open partition as a result of the mapping and the boundary should be countably n-rectifiable. Thus it appears that these two conditions are required at least. The choice of taking the interior points of the image fits naturally with the following examples. In Fig. 4.1, we have four domains as in (a), whose boundary is stationary but not minimizing. For grain boundary motion, (a) should move to (c) or a likewise figure, to have less length. If the four small triangles near the center are crushed to the line segment as indicated by arrows, and if the surrounding part is appropriately stretched so that the map is Lipschitz, we can obtain (c). Note that the image of these triangles will not be interior points. In Fig. 4.2, where both sides have the same labeling, we can pull up a portion of E_1 and cover the boundary. Then the covered part will be in the interior of the image of E_1, thus the resulting $\tilde{E}_1$ will be as shown in (c). This map is admissible, and this indicates that one can remove any "interior

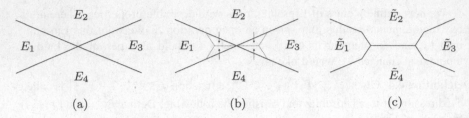

Fig. 4.1 Four different partitions

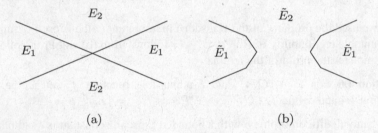

Fig. 4.2 Same kind of partition on both sides

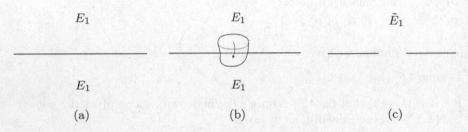

Fig. 4.3 Two different partitions

boundary" by a suitable admissible function. For Fig. 4.3 where there are only two kinds of labeling as shown, we can first use the map for Fig. 4.1, which produces some interior boundaries and then we can eliminate these interior boundaries just as we did for Fig. 4.2. Note that this is also a kind of non-diffeomorphic map which seems desirable when we have a picture like Fig. 4.3.

4.4   Restricted Class of Test Functions and Area–Reducing Admissible Functions

Up to this point, we defined a class of admissible functions which maps open partitions to itself. We next wish to find a way to choose an admissible function which reduces area. The idea is that we don't want to move any regular part (which should be moved by the mean curvature) while we want to "de-singularize" the kind

of singularities which are not measure minimizing. To do this, we first define a class
of test functions (and vector fields) as follows. Here, j will be used to introduce a
certain length scale subsequently.

Definition 4.7 For $j \in \mathbb{N}$, define

$$\mathcal{A}_j := \{\phi \in C^2(\mathbb{R}^{n+1}; \mathbb{R}^+) : \phi(x) \leq 1, |\nabla\phi(x)| \leq j\phi(x), \|\nabla^2\phi(x)\| \leq j\phi(x)\},$$

$$\mathcal{B}_j := \{g \in C^2(\mathbb{R}^{n+1}; \mathbb{R}^{n+1}) : |g(x)| \leq j, \|\nabla g(x)\| \leq j,$$

$$\|\nabla^2 g(x)\| \leq j, \|g\|_{L^2} \leq j\},$$

where the inequalities hold for all $x \in \mathbb{R}^{n+1}$.

It is easy to check that

$$\phi(x) \leq \phi(y) \exp(j|x - y|) \tag{4.6}$$

for any $\phi \in \mathcal{A}_j$ and $x, y \in \mathbb{R}^{n+1}$, so within a small distance of, say, $|x - y| \leq j^{-2}$,
ϕ behaves like a constant function. Any non-zero element of $\mathcal{A}_j$ is strictly positive
on $\mathbb{R}^{n+1}$, so ϕ cannot be a compactly supported function unless identically zero. On
the other hand, given any $\phi \in C_c^2(\mathbb{R}^{n+1}; \mathbb{R}^+)$ with $\phi < 1$, we can add an arbitrarily
small positive constant ε and one can check that $\phi + \varepsilon \in \mathcal{A}_j$ for all sufficiently
large j. Thus, in a sense, any element in $C_c^2(\mathbb{R}^{n+1}; \mathbb{R}^+)$ can be approximated
arbitrarily closely by functions in $\mathcal{A}_j$ for all large j, while functions in $\mathcal{A}_j$ behave
like constants on a small length scale of $o(j^{-1})$.

With this, we next restrict the admissible class as follows.

Definition 4.8 For $\mathcal{E} = \{E_i\}_{i=1}^N \in \mathcal{OP}^N$ and $j \in \mathbb{N}$, define $\mathbf{E}(\mathcal{E}, j)$ to be the set of
all $\mathcal{E}$-admissible functions $f : \mathbb{R}^{n+1} \to \mathbb{R}^{n+1}$ such that:

(a) $|f(x) - x| \leq j^{-2}$ for all $x \in \mathbb{R}^{n+1}$,
(b) $\mathcal{L}^{n+1}(\tilde{E}_i \Delta E_i) \leq j^{-1}$ for all $i = 1, \ldots, N$ and where $\{\tilde{E}_i\}_{i=1}^N = f_\star \mathcal{E}$,
(c) $\|\partial f_\star \mathcal{E}\|(\phi) \leq \|\partial \mathcal{E}\|(\phi)$ for all $\phi \in \mathcal{A}_j$.

The conditions (a) and (b) are clear in meaning: f is allowed to change little
in distance and in the Lebesgue measures of open partitions. The last condition
(c) is more subtle. Note that $\mathcal{A}_j$ contains the constant function $\phi = 1$ for all
$j \in \mathbb{N}$. Thus, to be in $\mathbf{E}(\mathcal{E}, j)$, $f_\star$ must be a measure–reducing map since
$\|\partial f_\star \mathcal{E}\|(1) = \mathcal{H}^n(\cup_{i=1}^N \partial \tilde{E}_i) \leq \|\partial \mathcal{E}\|(1) = \mathcal{H}^n(\cup_{i=1}^N \partial E_i)$. But this does not mean
that any measure–reducing map is in this class. Roughly speaking, unless $f_\star$ reduces
measure drastically (by some factor) within the length scale of j^{-2}, f cannot be in
$\mathbf{E}(\mathcal{E}, j)$. In fact, with this definition, $f \in \mathbf{E}(\mathcal{E}, j)$ cannot move a smooth part of the
boundary $\partial \mathcal{E}$ for all sufficiently large j. Just to see how this works, think about a
simple example of a circle of radius 1. As an area–reducing map, let us consider
f which shrinks this circle to the one with radius $1 - \varepsilon$ with $\varepsilon \leq j^{-2}$. Choose
$\phi(x) = 2(1 - |x|) + 1/2$ near $|x| = 1$ and otherwise truncate it smoothly to be

a positive constant. ϕ belongs to $\mathcal{A}_j$ for all sufficiently large j, as one can check easily. For such ϕ and circle shrinking f, (c) cannot be satisfied for any small $\varepsilon > 0$ since

$$\|\partial f_\star \mathcal{E}\|(\phi) - \|\partial \mathcal{E}\|(\phi) = \int_{\partial B_{1-\varepsilon}} \phi \, d\mathcal{H}^1 - \int_{\partial B_1} \phi \, d\mathcal{H}^1 = (3 - 4\varepsilon)\varepsilon\pi > 0.$$

This is not a proof, but one can argue by localizing this argument that f needs to be the identity map for all sufficiently large j around smooth points under the condition of (c). On the other hand, for the situation described in Fig. 4.1, if the size of this picture is smaller than j^{-2}, then the reduction of length by the described map is drastic enough to satisfy (c) since ϕ stays more or less like a constant on this scale. A more precise result on this is the following.

Lemma 4.9 ([7, 4.10]) *If* $\mathcal{E} = \{E_i\}_{i=1}^N \in \mathcal{OP}^N$, $j \in \mathbb{N}$, C *is a compact set of* $\mathbb{R}^{n+1}$, $f : \mathbb{R}^{n+1} \to \mathbb{R}^{n+1}$ *is a* $\mathcal{E}$-*admissible function such that:*

(a) $\{x : f(x) \neq x\} \cup \{f(x) : f(x) \neq x\} \subset C$,
(b) $|f(x) - x| \leq j^{-2}$ *for all* $x \in \mathbb{R}^{n+1}$,
(c) $\mathcal{L}^{n+1}(\tilde{E}_i \triangle E_i) \leq j^{-1}$ *for all* $i = 1, \ldots, N$ *and where* $\{\tilde{E}_i\}_{i=1}^N = f_\star \mathcal{E}$,
(d) $\|\partial f_\star \mathcal{E}\|(C) \leq \exp(-j \operatorname{diam} C)\|\partial \mathcal{E}\|(C)$.

Then we have $f \in \mathbf{E}(\mathcal{E}, j)$.

Proof By (a), $f(x) = x$ on $\mathbb{R}^{n+1} \setminus C$. For f to be in $\mathbf{E}(\mathcal{E}, j)$, we only need to check Definition 4.8(c). We have for $\phi \in \mathcal{A}_j$

$$\|\partial f_\star \mathcal{E}\|(\phi) - \|\partial \mathcal{E}\|(\phi) = \|\partial f_\star \mathcal{E}\| \llcorner_C (\phi) - \|\partial \mathcal{E}\| \llcorner_C (\phi)$$

$$\leq (\max_C \phi)\|\partial f_\star \mathcal{E}\|(C) - (\min_C \phi)\|\partial \mathcal{E}\|(C)$$

$$\leq \{(\max_C \phi) \exp(-j \operatorname{diam} C) - (\min_C \phi)\}\|\partial \mathcal{E}\|(C)$$

$$\leq 0,$$

where we used (c) and (4.6) which implies $\max_C \phi \leq \min_C \phi \exp(j \operatorname{diam} C)$. $\square$

Thus, if (a)–(c) are satisfied and if f reduces measure by a factor of, for example, $1/2$ within the ball of diameter j^{-2}, f belongs to $\mathbf{E}(\mathcal{E}, j)$. We then define a measure of how much one can reduce measure as follows.

Definition 4.10 For $\mathcal{E} \in \mathcal{OP}^N$ and $j \in \mathbb{N}$, we define

$$\Delta_j\|\partial \mathcal{E}\| := \Delta_j\|\partial \mathcal{E}\|(\mathbb{R}^{n+1}) := \inf_{f \in \mathbf{E}(\mathcal{E}, j)} \left(\|\partial f_\star \mathcal{E}\|(\mathbb{R}^{n+1}) - \|\partial \mathcal{E}\|(\mathbb{R}^{n+1})\right). \quad (4.7)$$

Since the identity map $f(x) = x$ always belongs to $\mathbf{E}(\mathcal{E}, j)$, we have $\Delta_j\|\partial \mathcal{E}\| \leq 0$. As we explained above, if $\partial \mathcal{E}$ is a smooth boundary, $\mathbf{E}(\mathcal{E}, j) = \{\mathrm{Id}\}$ and $\Delta_j\|\partial \mathcal{E}\| = 0$ for all sufficiently large j. Having $\Delta_j\|\partial \mathcal{E}\| = o(1)$ as $j \to \infty$ means roughly that the measure of $\partial\mathcal{E}$ cannot be reduced on a small scale of order j^{-2}. This helps us to control the behavior of measures on the small scale as $j \to \infty$.

4.5   Construction of Approximate Mean Curvature Flow

With the definitions up to this point, we can define a sequence of approximation, still without going too much into the detail. After we see how it is done, we later worry about what these maps do. Given $j \in \mathbb{N}$, in principle, we choose $\varepsilon_j \approx j^{-6} \ll j^{-2}$ and $\Delta t_j \approx \varepsilon_j^{3n+20} \ll \varepsilon_j$, where these choices are made due to various estimates detailed in [22]. The point of the choice is that ε_j, which is the length scale of smoothing of mean curvature vector, is much smaller than that of the measure minimizing scale of j^{-2}, and Δt_j is yet much smaller than ε_j. Then we define $\mathcal{E}_{j,k}$, $k \geq 0$, which is the approximate mean curvature flow at the time $k\Delta t_j$, as follows. In the end, we want to obtain some approximate Brakke flow where the test functions are in $\mathcal{A}_j$.

(1) Define $\mathcal{E}_{j,0} := \{E_{0,1}, \ldots, E_{0,N}\}$, where these sets appear at the beginning of the present section. Note that due to the assumption that $\Gamma_0 = \cup_{i=1}^N \partial E_{0,i}$ is closed, countably n-rectifiable and $\mathcal{H}^n(\Gamma_0) < \infty$, we have $\mathcal{E}_{j,0} \in \mathcal{OP}^N$.

(2) Assume inductively that $\mathcal{E}_{j,k-1} \in \mathcal{OP}^N$ is defined. Choose

$$f_{j,k} \in \mathbf{E}(\mathcal{E}_{j,k-1}, j)$$

such that

$$\|\partial(f_{j,k})_\star\mathcal{E}_{j,k-1}\|(\mathbb{R}^{n+1}) - \|\partial\mathcal{E}_{j,k-1}\|(\mathbb{R}^{n+1}) \leq (1 - j^{-5})\Delta_j\|\partial\mathcal{E}_{j,k-1}\|$$

$$(4.8)$$

and define

$$\mathcal{E}_{j,k}^* := (f_{j,k})_\star\mathcal{E}_{j,k-1}.$$

The inequality (4.8) means that we choose $f_{j,k}$ which is close to measure minimizing in $\mathbf{E}(\mathcal{E}_{j,k-1}, j)$. The factor $(1 - j^{-5})$ is somewhat arbitrary at this point. The minimizer of $\Delta_j\|\partial\mathcal{E}_{j,k-1}\|$ may not exist, so we choose something close to the infimum. Since $f_{j,k}$ is $\mathcal{E}_{j,k-1}$-admissible, we have $\mathcal{E}_{j,k}^* \in \mathcal{OP}^N$ and moreover,

$$\|\partial\mathcal{E}_{j,k}^*\|(\phi) \leq \|\partial\mathcal{E}_{j,k-1}\|(\phi) \text{ for all } \phi \in \mathcal{A}_j. \qquad (4.9)$$

Note also that when $\partial\mathcal{E}_{j,k-1}$ is smooth and j is sufficiently large, as explained in the previous section, $f_{j,k}$ is the identity map and $\mathcal{E}_{j,k}^* = \mathcal{E}_{j,k-1}$.

(3) For $\partial \mathcal{E}_{j,k}^*$, as in (4.3) and (4.4) where Γ_0 is replaced by $\partial \mathcal{E}_{j,k}^*$, compute the approximate mean curvature vectors $\tilde{h}_{\varepsilon_j}(\partial \mathcal{E}_{j,k}^*, y)$ and $h_{\varepsilon_j}(\partial \mathcal{E}_{j,k}^*, y)$. Define

$$\hat{f}_{j,k}(x) := x + \Delta t_j h_{\varepsilon_j}(\partial \mathcal{E}_{j,k}^*, x).$$

For the moment, let us assume that $\hat{f}_{j,k}$ is a diffeomorphism. Then define

$$\mathcal{E}_{j,k} := (\hat{f}_{j,k})_\star \mathcal{E}_{j,k}^*.$$

Since $\hat{f}_{j,k}$ is a diffeomorphism, $\mathcal{E}_{j,k} \in \mathcal{OP}^N$.

Thus, $\mathcal{E}_{j,k}$ can be inductively defined as long as $\hat{f}_{j,k}$ is a diffeomorphism. In the following, during the time interval of $[0, j]$ which is equivalent to k ranging over 1 to $[j/\Delta t_j]$ in terms of the number of inductive steps, we will see that $\|\partial \mathcal{E}_{j,k}^*\|(\mathbb{R}^{n+1})$ stays bounded by $\|\partial \mathcal{E}_{j,0}\|(\mathbb{R}^{n+1}) + 1 = \mathcal{H}^n(\Gamma_0) + 1$. The reason behind this is that we are constructing an approximation to the mean curvature flow which reduces measure in time, thus this is an expected property. Then by Lemma 4.2 with Γ_0 replaced by $\partial \mathcal{E}_{j,k}$, as long as $\Delta t_j \ll \varepsilon_j$, we can be sure that $\hat{f}_{j,k}$ is a diffeomorphism.

4.6   Estimates Related to h_ε and the Measure–Reducing Property

In the previous section, we saw two maps $f_{j,k}$ and $\hat{f}_{j,k}$. What the first map $f_{j,k}$ does to $\|\partial \mathcal{E}_{j,k-1}\|$ is simple in that it only reduces measure tested for any $\phi \in \mathcal{A}_j$ including $\phi = 1$, as in (4.9). We now look into what $\hat{f}_{j,k}$ does. For simplicity, let us denote $\partial \mathcal{E}_{j,k}^*$ by Γ and also we drop all indices such as j and k in the next two lemmas.

Lemma 4.11 *Suppose* $h : \mathbb{R}^{n+1} \to \mathbb{R}^{n+1}$ *is smooth,* $\hat{f}(x) := x + h(x)\Delta t$ *and* $\hat{\Gamma} := \hat{f}(\Gamma)$. *Then we have*

$$\left|\mathcal{H}^n(\hat{\Gamma}) - \mathcal{H}^n(\Gamma) - \delta|\Gamma|(h\Delta t)\right| \leq c(n)\mathcal{H}^n(\Gamma)(\sup|\nabla h|\Delta t)^2. \tag{4.10}$$

Proof By the change of variables (or more precisely the area formula for countably n-rectifiable set Γ), we have

$$\mathcal{H}^n(\hat{\Gamma}) = \int_\Gamma |\Lambda_n \nabla \hat{f}(x) \circ \mathrm{T}_x \Gamma| \, \mathrm{d}\mathcal{H}^n(x),$$

where the integrand is the Jacobian of the map $\nabla \hat{f}(x) \llcorner_{T_x \Gamma} : \hat{T}_x \Gamma \to \mathbb{R}^{n+1}$. Namely, fix $x \in \Gamma$ where Γ has the approximate tangent space $T_x \Gamma$ and let $e_1, \ldots, e_n$ be an orthonormal basis for $T_x \Gamma$. Then we have

$$|\Lambda_n \nabla \hat{f}(x) \circ T_x \Gamma|^2 = \det \left((\nabla_{e_i} \hat{f}(x))^\top \circ (\nabla_{e_j} \hat{f}(x)) \right)_{1 \leq i, j \leq n}.$$

Since $\nabla_{e_i} \hat{f}(x) = e_i + \nabla_{e_i} h(x) \Delta t$, the expansion of $\mathcal{H}^n(\hat{\Gamma})$ with respect to Δt may be computed explicitly. The zeroth–order term is $\mathcal{H}^n(\Gamma)$. The term $\delta|\Gamma|(h \Delta t)$ is in fact the first–order expansion and may be recognized using (1.12) as

$$\delta|\Gamma|(h \Delta t) = \Delta t \int_\Gamma \text{div}_{T_x} \Gamma h(x) \, d\mathcal{H}^n(x).$$

In sum, since we have

$$\left| |\Lambda_n \nabla \hat{f}(x) \circ T_x \Gamma| - 1 - \Delta t \, \text{div}_{T_x} \Gamma h \right| \leq c(n)(|\nabla h(x)| \Delta t)^2,$$

this proves (4.10). □

Note that we will have a uniform bound on $\mathcal{H}^n(\Gamma)$ which gives also an estimate on $\varepsilon^4 |\nabla h_\varepsilon(\Gamma, x)|$ by Lemma 4.2. Since the right-hand side is quadratic in Δt and $\Delta t \ll \varepsilon$, we see that (4.10) is less than, for example, $\varepsilon^4 \Delta t$.

Lemma 4.12

$$\delta|\Gamma|(h_\varepsilon(\Gamma, \cdot)) = - \int_{\mathbb{R}^{n+1}} |\tilde{h}_\varepsilon(\Gamma, x)|^2 (\Phi_\varepsilon * \mathcal{H}^n \llcorner_\Gamma (x) + \varepsilon) \, dx, \tag{4.11}$$

where we define $\Phi_\varepsilon * \mathcal{H}^n \llcorner_\Gamma (x) = \int_\Gamma \Phi_\varepsilon (x - y) \, d\mathcal{H}^n(x)$.

Proof Write $\tilde{h}_\varepsilon(\Gamma, y)$ as $\tilde{h}_\varepsilon(y)$ or $\tilde{h}_\varepsilon$ and likewise for $h_\varepsilon(\Gamma, y)$. By (4.4),

$$\delta|\Gamma|(h_\varepsilon) = \int_\Gamma T_x \Gamma \cdot \nabla h_\varepsilon(x) \, d\mathcal{H}^n(x)$$

$$= \int_\Gamma \int_{\mathbb{R}^{n+1}} T_x \Gamma \cdot \nabla \Phi_\varepsilon (x - y) \otimes \tilde{h}_\varepsilon(y) \, dy d\mathcal{H}^n(x).$$

By (4.3),

$$\int_{\mathbb{R}^{n+1}} |\tilde{h}_\varepsilon(y)|^2 (\Phi_\varepsilon * \mathcal{H}^n \llcorner_\Gamma (y) + \varepsilon) \, dy$$

$$= - \int_{\mathbb{R}^{n+1}} \tilde{h}_\varepsilon(y) \cdot \int_\Gamma (T_x \Gamma) \circ \nabla \Phi_\varepsilon (x - y) \, d\mathcal{H}^n(x) dy.$$

Since $\tilde{h}_\varepsilon(y) \cdot (T_x \dot{\,}\Gamma) \circ \nabla \Phi_\varepsilon(x - y) = T_x \Gamma \cdot \nabla \Phi_\varepsilon(x - y) \otimes \tilde{h}_\varepsilon(y)$ and by the Fubini theorem, we have (4.11). □

Now combining these two lemmas, we obtain by recovering the indices

$$
\|\partial \mathcal{E}_{j,k}\|(\mathbb{R}^{n+1}) - \|\partial \mathcal{E}^*_{j,k}\|(\mathbb{R}^{n+1})
$$
$$
+ \Delta t_j \int_{\mathbb{R}^{n+1}} |\tilde{h}_{\varepsilon_j}(\partial \mathcal{E}^*_{j,k}, x)|^2 (\Phi_{\varepsilon_j} * \mathcal{H}^n \llcorner_{\partial \mathcal{E}^*_{j,k}}(x) + \varepsilon_j) \, \mathrm{d}x \qquad (4.12)
$$
$$
\leq c(n)\|\partial \mathcal{E}^*_{j,k}\|(\mathbb{R}^{n+1})(\sup |\nabla h_{\varepsilon_j}(\partial \mathcal{E}^*_{j,k}, x)|\Delta t_j)^2.
$$

Recalling that we have (4.8), we obtain from above the following.

Lemma 4.13

$$
\|\partial \mathcal{E}_{j,k}\|(\mathbb{R}^{n+1}) - \|\partial \mathcal{E}_{j,k-1}\|(\mathbb{R}^{n+1})
$$
$$
+ \Delta t_j \int_{\mathbb{R}^{n+1}} |\tilde{h}_{\varepsilon_j}(\partial \mathcal{E}^*_{j,k}, x)|^2 (\Phi_{\varepsilon_j} * \mathcal{H}^n \llcorner_{\partial \mathcal{E}^*_{j,k}}(x) + \varepsilon_j) \, \mathrm{d}x \qquad (4.13)
$$
$$
- (1 - j^{-5})\Delta_j\|\partial \mathcal{E}_{j,k-1}\|
$$
$$
\leq c(n)\|\partial \mathcal{E}_{j,k-1}\|(\mathbb{R}^{n+1})(\sup |\nabla h_{\varepsilon_j}(\partial \mathcal{E}^*_{j,k}, x)|\Delta t_j)^2.
$$

Note that the 3rd and 4th terms are non-negative. If we inductively assume up to $k - 1$ that $\|\partial \mathcal{E}_{j,k-1}\|(\mathbb{R}^{n+1}) \leq \|\partial \mathcal{E}_{j,0}\|(\mathbb{R}^{n+1}) + 1 =: M$, by Lemma 4.2, we have (recall also $\|\partial \mathcal{E}^*_{j,k}\|(\mathbb{R}^{n+1}) \leq \|\partial \mathcal{E}_{j,k-1}\|(\mathbb{R}^{n+1})$)

$$
c(n)\|\partial \mathcal{E}_{j,k-1}\|(\mathbb{R}^{n+1})(\sup |\nabla h_{\varepsilon_j}(\partial \mathcal{E}_{j,k}, x)|\Delta t_j)^2 \leq c(n, M)(\varepsilon_j^{-4}\Delta t_j)^2.
$$

Thus adding (4.13) all the way up to $k \leq j/\Delta t_j$, we have

$$
\|\partial \mathcal{E}_{j,k}\|(\mathbb{R}^{n+1}) - \|\partial \mathcal{E}_{j,0}\|(\mathbb{R}^{n+1}) \leq c(n, M)\varepsilon_j^{-8} j \Delta t_j
$$

and as long as $\Delta t_j \ll \varepsilon_j$ (in fact, as mentioned before, $\varepsilon_j \approx j^{-6}$ and $\Delta t_j \approx \varepsilon_j^{3n+20}$ in [22]), we have the right-hand side smaller than ε_j for all sufficiently large j, for example. Thus we have $\|\partial \mathcal{E}_{j,k}\|(\mathbb{R}^{n+1}) \leq M$, closing the inductive step. In particular, from the last inequality, we obtain:

Lemma 4.14 *For all $k = 1, \ldots, [j/\Delta t_j]$, we have*

$$
\|\partial \mathcal{E}_{j,k}\|(\mathbb{R}^{n+1}) \leq \|\partial \mathcal{E}_0\|(\mathbb{R}^{n+1}) + \varepsilon_j.
$$

This inductive argument combined with (4.13) gives also the following estimate for j sufficiently large and for all $k = 1, \ldots, [j/\Delta t_j]$:

$$\|\partial \mathcal{E}_{j,k}\|(\mathbb{R}^{n+1}) - \|\partial \mathcal{E}_{j,k-1}\|(\mathbb{R}^{n+1})$$

$$+ \Delta t_j \int_{\mathbb{R}^{n+1}} |\tilde{h}_{\varepsilon_j}(\partial \mathcal{E}_{j,k}^*, x)|^2 (\Phi_{\varepsilon_j} * \mathcal{H}^n \llcorner_{\partial \mathcal{E}_{j,k}^*}(x) + \varepsilon_j)\, dx \qquad (4.14)$$

$$- (1 - j^{-5})\Delta_j \|\partial \mathcal{E}_{j,k-1}\| \leq c(n, M)\varepsilon_j \Delta t_j.$$

In the following, we want to define the discrete flow in terms of $\mathcal{E}_{j,k}$, and we want to do away with $\mathcal{E}_{j,k}^*$ in (4.14). $\mathcal{E}_{j,k}$ is obtained from $\mathcal{E}_{j,k}^*$ by $\hat{f}_{j,k}$ which deviates from the identity map by $h_{\varepsilon_j} \Delta t_j$ and Δt_j is $\ll \varepsilon_j$. In fact, by carrying out somewhat cumbersome estimates, we can prove (see [22, Proposition 5.7])

$$\left| \int_{\mathbb{R}^{n+1}} |\tilde{h}_{\varepsilon_j}(\partial \mathcal{E}_{j,k}^*, x)|^2 (\Phi_{\varepsilon_j} * \mathcal{H}^n \llcorner_{\partial \mathcal{E}_{j,k}^*}(x) + \varepsilon_j)\, dx \right. $$

$$\left. - \int_{\mathbb{R}^{n+1}} |\tilde{h}_{\varepsilon_j}(\partial \mathcal{E}_{j,k}, x)|^2 (\Phi_{\varepsilon_j} * \mathcal{H}^n \llcorner_{\partial \mathcal{E}_{j,k}}(x) + \varepsilon_j)\, dx \right| \leq \varepsilon_j. \qquad (4.15)$$

Thus (4.14) and (4.15) give:

Lemma 4.15 *For all sufficiently large j and for all $k = 1, \ldots, [j/\Delta t_j]$, we have*

$$\frac{\|\partial \mathcal{E}_{j,k}\|(\mathbb{R}^{n+1}) - \|\partial \mathcal{E}_{j,k-1}\|(\mathbb{R}^{n+1})}{\Delta t_j}$$

$$+ \int_{\mathbb{R}^{n+1}} |\tilde{h}_{\varepsilon_j}(\partial \mathcal{E}_{j,k}, x)|^2 (\Phi_{\varepsilon_j} * \mathcal{H}^n \llcorner_{\partial \mathcal{E}_{j,k}}(x) + \varepsilon_j)\, dx \qquad (4.16)$$

$$- (1 - j^{-5})\frac{\Delta_j \|\partial \mathcal{E}_{j,k-1}\|}{\Delta t_j} \leq c(n, M)\varepsilon_j + '\varepsilon_j.$$

Formally, (4.16) is an analogue of

$$\frac{d}{dt} \mathcal{H}^n(\Gamma(t)) \leq - \int_{\Gamma(t)} |h(\Gamma(t), x)|^2 \, d\mathcal{H}^n(x)$$

for the time interval of $[0, j]$ since each time step corresponds to Δt_j.

Though the argument up to this point shows that $\partial \mathcal{E}_{j,k}$ has an "almost" measure–reducing property in time, this is not certainly enough to prove that the limit is a Brakke flow. We need to see what $\hat{f}_{j,k}$ does more closely, namely, we need to investigate how $\hat{f}_{j,k}$ affects $\|\partial \mathcal{E}_{k,j}\|(\phi)$ for each $\phi \in \mathcal{A}_j$. To do so, we need to obtain ϕ-weighted versions of Lemma 4.15. Note that we have investigated the case of $\phi = 1$ so far. Without going into the detail, which is rather tedious, we may obtain an analogue of (4.16) as follows.

Lemma 4.16 ([22, Proposition 6.1]) *With the choices of ε_j, Δt_j as above and for all sufficiently large j, $k = 1 \ldots, [j/\Delta t_j]$ and $\phi \in \mathcal{A}_j$, we have*

$$\frac{\|\partial \mathcal{E}_{j,k}\|(\phi) - \|\partial \mathcal{E}_{j,k-1}\|(\phi)}{\Delta t_j} - \delta(\partial \mathcal{E}_{j,k}, \phi)(h_{\varepsilon_j}(\partial \mathcal{E}_{j,k}, \cdot)) \le \varepsilon_j^{\frac{1}{8}}. \tag{4.17}$$

Here, recall the definition (1.20). This is a time-discrete analogue of

$$\frac{d}{dt} \int_{\Gamma(t)} \phi(x) \, d\mathcal{H}^n(x) \le \int_{\Gamma(t)} -\phi(x)|h(\Gamma(t), x)|^2 + \nabla\phi(x) \cdot h(\Gamma(t), x) \, d\mathcal{H}^n(x)$$

due to the smooth case (1.22). Some other important ingredients are the ϕ-weighted version of Lemma 4.12 as follows.

Lemma 4.17 ([22, Proposition 5.4]) *Depending only on M and n, for all sufficiently large j, we have the following. Suppose $\mathcal{H}^k(\Gamma) \le M$ and $\phi \in \mathcal{A}_j$, then we have*

$$\left| \delta|\Gamma|(\phi h_{\varepsilon_j}(\Gamma, \cdot)) + \int_{\mathbb{R}^{n+1}} \phi|\tilde{h}_{\varepsilon_j}(\Gamma, x)|^2 (\Phi_{\varepsilon_j} * \mathcal{H}^n \llcorner_\Gamma (x) + \varepsilon_j) \, dx \right|$$
$$\le \varepsilon_j^{\frac{1}{4}} \left(\int_{\mathbb{R}^{n+1}} \phi|\tilde{h}_{\varepsilon_j}(\Gamma, x)|^2 (\Phi_{\varepsilon_j} * \mathcal{H}^n \llcorner_\Gamma (x) + \varepsilon_j) \, dx + 1 \right) \tag{4.18}$$

and

$$\int_\Gamma |h_{\varepsilon_j}(\Gamma, \cdot)|^2 \phi \, d\mathcal{H}^n$$
$$\le (1 + \varepsilon_j^{\frac{1}{2}}) \int_{\mathbb{R}^{n+1}} \phi|\tilde{h}_{\varepsilon_j}(\Gamma, x)|^2 (\Phi_{\varepsilon_j} * \mathcal{H}^n \llcorner_\Gamma (x) + \varepsilon_j) \, dx + \varepsilon_j^{\frac{1}{4}}. \tag{4.19}$$

When $\phi = 1$, the left-hand side of (4.18) is actually 0 as seen in Lemma 4.12. It is interesting to point out that Φ_ε being the Gaussian function (plus a truncation) is surprisingly essential to obtain (4.18). In fact, what we need is that $x\Phi_\varepsilon(x) + \varepsilon^2 \nabla \Phi_\varepsilon(x) = \varepsilon^2 c(\varepsilon)\nabla\psi(x)\hat{\Phi}(x)$ and that this is exponentially small as $\varepsilon \to 0$. The computation involves an interesting integration by parts calculation where this property is used. For me, this is a hidden mystery which makes the method work.

For $\delta(\partial \mathcal{E}_{j,k}, \phi)(h_{\varepsilon_j}(\partial \mathcal{E}_{j,k}, \cdot))$ in (4.17), writing $|\Gamma| = \partial \mathcal{E}_{j,k}$, we use (1.21) to compute (write also $h_{\varepsilon_j} = h_{\varepsilon_j}(|\Gamma|, \cdot)$)

$$\delta(|\Gamma|, \phi)(h_{\varepsilon_j}) = \delta|\Gamma|(\phi h_{\varepsilon_j}) + \int_{\mathbf{G}_n(\mathbb{R}^{n+1})} h_{\varepsilon_j} \cdot S^\perp(\nabla\phi) \, d|\Gamma|$$
$$\le \delta|\Gamma|(\phi h_{\varepsilon_j}) + \frac{1}{2} \int_\Gamma |h_{\varepsilon_j}|^2 \phi + \frac{|\nabla\phi|^2}{\phi} \, d\mathcal{H}^n \tag{4.20}$$

where we used the Cauchy–Schwarz inequality and $|S^{\perp}(\nabla\phi)| \leq |\nabla\phi|$. Using (4.18) and (4.19) in (4.20), we obtain

$$\delta(|\Gamma|, \phi)(h_{\varepsilon_j}) \leq 2\varepsilon_j^{\frac{1}{4}} + \frac{1}{2} \int_{\Gamma} \frac{|\nabla\phi|^2}{\phi} \, d\mathcal{H}^n \qquad (4.21)$$

for $\phi \in \mathcal{A}_j$.

4.7 Taking a Limit

For each $j \in \mathbb{N}$, we construct a sequence of open partitions $\{\mathcal{E}_{j,k}\}$ with $k = 1, \ldots, [j/\Delta t_j]$ and some desired estimates such as (4.16) and (4.17). It is natural to define a piece-wise constant approximate flow $\{\mathcal{E}_j(t)\}_{0 \leq t < j}$ from this discrete flow by

$$\mathcal{E}_j(t) := \mathcal{E}_{j,k} \text{ if } t \in ((k-1)\Delta t_j, k\Delta t_j] \qquad (4.22)$$

for $t \in [0, j]$. We next see that there exists a subsequence $\{j_l\}_{l=1}^{\infty}$ such that $\|\partial\mathcal{E}_{j_l}(t)\|$ converges to some Radon measure μ_t for all $t \geq 0$. It is important to emphasize that the subsequence does not depend on t and the proof resembles that for the compactness theorem for Brakke flows. We choose $p_j \in \mathbb{N}$ so that $\Delta t_j = 1/2^{p_j} \in (\varepsilon_j^{3n+20}/2, \varepsilon_j^{3n+20}]$. For $t \in [0, j]$, we have a uniform bound for $\|\partial\mathcal{E}_j(t)\|(\mathbb{R}^{n+1})$. Let $2_{\mathbb{Q}}$ be the set of all non-negative rational numbers of the form $k/2^l, k, l \in \mathbb{N}$. By the weak compactness of Radon measures and by a diagonal argument, there exists a subsequence j_l such that $\|\partial\mathcal{E}_{j_l}(t)\|$ converges to a Radon measure μ_t for all $t \in 2_{\mathbb{Q}}$ as $l \to \infty$. We fix this subsequence and prove that $\|\partial\mathcal{E}_{j_l}(t)\|$ converges for all $t \geq 0$. For this, let $\{\phi_q\}_{q=1}^{\infty}$ be a sequence in $C_c^2(\mathbb{R}^{n+1}; \mathbb{R}^+)$ which is dense in $C_c(\mathbb{R}^{n+1}; \mathbb{R}^+)$ with respect to the supremum norm. It suffices to see the convergence property of $\|\partial\mathcal{E}_{j_l}(t)\|(\phi_q)$. Define

$$g_q(t) := \mu_t(\phi_q) - t\|\nabla^2\phi_q\|_{\infty}\|\partial\mathcal{E}_0\|(\mathbb{R}^{n+1})$$

for $t \in 2_{\mathbb{Q}}$. Then we have:

Lemma 4.18 g_q is a monotone decreasing function on $2_{\mathbb{Q}}$.

Proof Since the claim is linear with respect to ϕ_q, we may assume that $\phi_q < 1$. Take any $0 < t_1 < t_2$ such that $t_1, t_2 \in 2_{\mathbb{Q}}$. Choose large enough l so that t_1 and t_2 are integer multiples of $1/2^{p_{j_l}} = \Delta t_{j_l}$ and that $t_2 < j_l$. We also fix an arbitrarily large $i \in \mathbb{N}$. Corresponding to this i, if necessary, choose a larger j_l so that $\phi_q + i^{-1} \in \mathcal{A}_{j_l}$. Recall that any non-zero element of $\mathcal{A}_j$ is strictly positive. Then using (4.17) and (4.21) and summing over $k = t_1/\Delta t_{j_l}, \ldots, t_2/\Delta t_{j_l}$, we have

$$\|\partial \mathcal{E}_{j_l}(t_2)\|(\phi_q + i^{-1}) - \|\partial \mathcal{E}_{j_l}(t_1)\|(\phi_q + i^{-1})$$

$$\leq 3\varepsilon_{j_l}^{\frac{1}{8}}(t_2 - t_1) + \frac{1}{2}\int_{t_1}^{t_2}\int_{\partial \mathcal{E}_{j_l}(t)}\frac{|\nabla \phi_q|^2}{\phi_q + i^{-1}}\,d\mathcal{H}^n dt.$$

The $\liminf$ of the left-hand side as $l \to \infty$ is $\geq \mu_{t_2}(\phi_q) - \mu_{t_1}(\phi_q) - i^{-1}\|\partial \mathcal{E}_0\|(\mathbb{R}^{n+1})$. Since $|\nabla \phi_q|^2/\phi_q \leq 2\|\nabla^2 \phi_q\|_\infty$ and by Lemma 4.14, we obtain

$$\mu_{t_2}(\phi_q) - \mu_{t_1}(\phi_q) - i^{-1}\|\partial \mathcal{E}_0\|(\mathbb{R}^{n+1}) \leq (t_2 - t_1)\|\nabla^2 \phi_q\|_\infty \|\partial \mathcal{E}_0\|(\mathbb{R}^{n+1}).$$

Since i is arbitrary, we may conclude that g_q is monotone decreasing on $2_\mathbb{Q}$. $\qquad\square$

Since g_q is monotone decreasing on $2_\mathbb{Q}$, for each $x \in \mathbb{R}^+$, there exist both the right and left limit of g_q at x. By monotonicity again, it is discontinuous only on a countable set. Let $D \subset [0, \infty)$ be a set of points where g_q is discontinuous for some $q \in \mathbb{N}$ which is countable. For $t \in [0, \infty) \setminus D$, by continuity, $\mu_t(\phi_q)$ is well-defined. It is a little exercise to prove that $\|\partial \mathcal{E}_{j_l}(t)\|(\phi_q)$ converges to $\mu_t(\phi_q)$ for all $t \in [0, \infty) \setminus D$. For $t \in D$, since D is countable, we may choose a further subsequence so that $\|\partial \mathcal{E}_{j_l}(t)\|$ converges to a Radon measure μ_t for such t.

So far, we proved that there is a limiting Radon measure μ_t for each $t \geq 0$, which should be the surface measures of moving hypersurfaces. On the other hand, note that it is not clear at this point what μ_t is beyond the mere fact that it is a Radon measure. We even do not know if μ_t is an "n-dimensional measure" in the sense that μ_t may be locally like $\mathcal{L}^{n+1}$. To prove that the limit constitutes a Brakke flow, we need to show that μ_t is a measure of specific type, namely, we want it to be $\|V_t\|$ for some $V_t \in \mathbf{IV}_n(\mathbb{R}^{n+1})$ for almost all times.

4.8   Compactness Theorems and Last Steps

Here we see what are the controlled quantities for the converging sequence $\{\partial \mathcal{E}_{j_l}(t)\}_{l=1}^\infty$. From Lemma 4.15 and recalling (4.22), we have

$$\|\partial \mathcal{E}_j(t)\|(\mathbb{R}^{n+1}) - \|\partial \mathcal{E}_j(0)\|(\mathbb{R}^{n+1})$$

$$+ \int_0^t \int_{\mathbb{R}^{n+1}} |\tilde{h}_{\varepsilon_j}(\partial \mathcal{E}_j(s), x)|^2 (\Phi_{\varepsilon_j} * \mathcal{H}^n \llcorner_{\partial \mathcal{E}_j(s)}(x) + \varepsilon_j)\,dxds$$

$$- (1 - j^{-5})\int_0^t \frac{\Delta_j\|\partial \mathcal{E}_j(s)\|}{\Delta t_j}\,ds \leq (c(n, M)\varepsilon_j + \varepsilon_j)t$$

for all $0 \leq t \leq j$ with $t/\Delta t_j \in \mathbb{N}$. In particular, this gives two inequalities

$$\limsup_{j\to\infty} \int_0^j \int_{\mathbb{R}^{n+1}} |\tilde{h}_{\varepsilon_j}(\partial \mathcal{E}_j(s), x)|^2 (\Phi_{\varepsilon_j} * \mathcal{H}^n \llcorner_{\partial \mathcal{E}_j(s)}(x) + \varepsilon_j)\,dxds$$

$$\leq \|\partial \mathcal{E}_0\|(\mathbb{R}^{n+1})$$

$$(4.23)$$

and

$$\limsup_{j\to\infty} \int_0^j \frac{|\varDelta_j| \|\partial\mathcal{E}_j(s)\|}{\varDelta t_j}\, ds \le \|\partial\mathcal{E}_0\|(\mathbb{R}^{n+1}). \tag{4.24}$$

These inequalities hold also for the subsequence $\{\partial\mathcal{E}_{j_l}(t)\}_{l=1}^\infty$. From (4.23) and Fatou's Lemma, we have

$$\liminf_{l\to\infty} \int_{\mathbb{R}^{n+1}} |\tilde{h}_{\varepsilon_{j_l}}(\partial\mathcal{E}_{j_l}(t), x)|^2 (\varPhi_{\varepsilon_{j_l}} * \mathcal{H}^n \llcorner_{\partial\mathcal{E}_{j_l}(t)}(x) + \varepsilon_{j_l})\, dx < \infty \tag{4.25}$$

for almost all $t \in \mathbb{R}^+$. Also, for any sequence of positive numbers $\{\alpha_l\}_{l=1}^\infty$ such that $\lim_{l\to\infty} \varDelta t_{j_l}/\alpha_l = 0$, we have from (4.24)

$$\lim_{l\to\infty} \int_0^{j_l} \frac{|\varDelta_{j_l}| \|\partial\mathcal{E}_{j_l}(s)\|}{\alpha_l}\, ds = 0.$$

Thus possibly choosing a further subsequence, we have

$$\lim_{l\to\infty} \frac{|\varDelta_{j_l}| \|\partial\mathcal{E}_{j_l}(t)\|}{\alpha_l} = 0 \tag{4.26}$$

for almost all $t \in \mathbb{R}^+$. We focus on $t \in \mathbb{R}^+$ such that both (4.25) and (4.26) are satisfied, which holds for a.e. t. Recall that $\{\partial\mathcal{E}_{j_l}(t)\}_{l=1}^\infty$ is a sequence of unit density integral varifolds. We know already that as a sequence of Radon measures on $\mathbb{R}^{n+1}$, it converges to μ_t. Considering the compactness theorem of integral varifolds, Theorem 1.15, we naturally wish to claim that μ_t is also integral. Now the difficulty here is that we only have the L^2 control of approximate mean curvature vectors $\tilde{h}_{\varepsilon_{j_l}}$, not the "real" mean curvature vectors. If we had control of the L^2 norm of mean curvature vectors instead, by Theorem 1.15, we could conclude that the limit varifold is in fact integral. But here, due to the smoothing by $\varPhi_{\varepsilon_{j_l}}$, we do not see what is happening on a scale smaller than ε_{j_l}. To compensate this lack of resolution, we have (4.26) which tells us that $\partial\mathcal{E}_{j_l}(t)$ is "almost measure–minimizing" on a length scale of j_l^{-2}, which is much larger than ε_{j_l} since we take $\varepsilon_j \approx j^{-6}$.

In [22, Section 7, 8], we prove that there exists $V_t \in \mathbf{IV}_n(\mathbb{R}^{n+1})$ such that $\mu_t = \|V_t\|$. More precisely, with $\alpha_l = 1$, we first prove that V_t is a rectifiable varifold, and also by assuming stronger convergence with $\alpha_l = 1/j_l^{2(n+1)}$ (which we can do since $\varDelta t_j \approx \varepsilon_j^{3n+20}$ and $\varepsilon_j \approx j^{-6}$), we prove that V_t is integral. The proof is rather long and technical and I do not intend to repeat it here, but I shall just describe intuitively what is involved.

The first step is to prove rectifiability of the limit varifold. By the compactness of Radon measures, there is some converging subsequence such that the integral quantities in (4.25) are uniformly bounded in addition to (4.26). Let V_t be any such limit. By the lower-semicontinuity property and using (4.19), we can show

$$\int_{\mathbb{R}^{n+1}} |h(V_t, \cdot)|^2 \, d\|V_t\|$$

$$\leq \liminf_{l \to \infty} \int_{\mathbb{R}^{n+1}} |\tilde{h}_{\varepsilon_{j_l}}(\partial \mathcal{E}_{j_l}(t), x)|^2 (\Phi_{\varepsilon_{j_l}} * \mathcal{H}^n \llcorner_{\partial \mathcal{E}_{j_l}(t)}(x) + \varepsilon_{j_l}) \, dx < \infty.$$

So in particular the first variation of V_t is bounded. By Theorem 1.17, if we can show a positive lower density bound for V_t, we may conclude that V_t is rectifiable. The logic for having this lower density bound is as follows. Suppose that there are a lot of points $\{x_k\}$ where they are part of the support of $\|V_t\|$, but $\|V_t\|(B_r(x_k)) = o(r^n)$. Then one can arrange these points so that the similar smallness holds for $\|\partial \mathcal{E}_{j_l}(t)\|$ for $r \leq j^{-2}$. One can prove that there is a threshold value c depending on n such that, if $\|\partial \mathcal{E}_{j_l}(t)\|(B_r(x_k)) \leq cr^n$, then there is an admissible Lipschitz map which reduces the mass by the factor of $1/2$ in $B_r(x_k)$. If one has too many such balls, then it means that one can reduce the mass of $\partial \mathcal{E}_{j_l}(t)$ by an admissible Lipschitz map for all large l and we get a contradiction to (4.26) with $\alpha_l = 1$.

Once we establish the rectifiability of a limit V_t (which may in principle depend on the choice of subsequence of j_l), we know that $\|V_t\| = \theta \mathcal{H}^n \llcorner \Gamma$ for some positive function θ and a countably rectifiable set Γ. Since $\|V_t\| = \lim_{l \to \infty} \|\partial \mathcal{E}_{j_l}(t)\| = \mu_t$, we have $\mu_t = \theta \mathcal{H}^n \llcorner \Gamma$, which shows that μ_t is an n-dimensional measure supported on a countably rectifiable set. Also θ and Γ are uniquely determined since μ_t is already fixed. Since V_t is determined completely by θ and Γ, any limit varifold obtained as above has to be the same.

The next step is to show V_t is integral, which boils down to showing that the multiplicity θ is integer valued a.e. Since we know V_t is rectifiable, we can focus on a generic point where the approximate tangent space of V_t exists. Let such a point be the origin, and let the approximate tangent space be $\mathbb{R}^n \times \{0\} \subset \mathbb{R}^{n+1}$. We can also assume that the origin is a Lebesgue point for θ and that $\lim_{l \to \infty} (\tau_{r_l})_\sharp V_t = \theta(0)|\mathbb{R}^n \times \{0\}|$ for any $r_l \to 0+$. We can choose r_l converging very slowly so that $\lim_{l \to \infty} (\tau_{r_l})_\sharp \partial \mathcal{E}_{j_l}(t) = \theta(0)|\mathbb{R}^n \times \{0\}|$ as well. Then we show that $(\tau_{r_l})_\sharp \partial \mathcal{E}_{j_l}(t)$ looks more or less like a finite number of n-dimensional sheets in terms of measure which are almost parallel to $\mathbb{R}^n \times \{0\}$. One uses a monotonicity type estimate for length scales larger than j^{-2}, and for smaller scales, we use the minimizing property to show that the "sheeted picture" is correct on a small scale of j^{-2}. This proves that $\theta(0) \in \mathbb{N}$ in the end.

Once we show that μ_t corresponds to $\|V_t\|$ for some $V_t \in \mathbf{IV}_n(\mathbb{R}^{n+1})$ for a.e. $t \in \mathbb{R}^+$, we still need to prove that this V_t satisfies (2.11), and also that there exists a set of domains $E_i(t)$ moving continuously in time and so that the boundaries coincide with the Brakke flow. Verifying (2.11) corresponds to showing that the inequality (4.17) converges to (2.11), which is carried out in [22, Section 9]. To show the continuity of domains, we use Huisken's monotonicity formula in an essential way.

It is interesting that Huisken's monotonicity formula is not used until this step, at least as far as our existence results are concerned.

4.9 Comments on the Existence Results

Though the existence results are satisfactory for the generality of initial data and for being time-global, it is desirable if we know more on the behaviors. For example, even though we have the specific Lipschitz map in the construction of time-discrete flow, we do not know if there is any extra condition that V_t satisfies in general, other than the fact that it is a Brakke flow. I suspect that there are more things one can say about the flow obtained in [22]. To put the question somewhat differently in a simpler situation, consider the following question. Suppose that we have an open partition of N elements $\mathcal{E}$ such that $\partial\mathcal{E}$ is a singular stationary varifold in $\mathbb{R}^{n+1}$. Since any stationary varifold in $\mathbb{R}^{n+1}$ cannot be compact, we cannot confine ourselves to a finite measure case. So let us assume that $\|\partial\mathcal{E}\|(B_R) \leq c\exp(cR)$ for some $c > 0$. With this as the initial data, [22] gives the time-global existence of a Brakke flow starting from $\partial\mathcal{E}$. We may ask if there is some condition on $\partial\mathcal{E}$ such that the obtained Brakke flow through our time-discrete construction is necessarily time-independent. Let us define somewhat loosely the following.

Definition 4.19 A unit density varifold $\partial\mathcal{E}$ with $\mathcal{E} \in \mathcal{OP}_N$ and with at most exponential mass growth is called dynamically stable if any Brakke flow starting from $\partial\mathcal{E}$ via the construction of [22] is time-independent.

For example, a single triple junction should be measure minimizing under the Lipschitz map, so it should be dynamically stable. On the other hand, two crossing lines should move since we may reduce the mass by a suitable Lipschitz map. Is there any general characterization differentiating the two? For $n = 1$, we already know from [2] that spt $\|\partial\mathcal{E}\|$ consists of locally finite line segments with their end points meeting at junctions in a way that they are stationary. With this simple structure, the answer for $n = 1$ should be that the support of dynamically stable varifold consists of straight line segments jointed by triple junctions, even though the rigorous proof is not written down. For $n > 1$, it is reasonable to expect that a dynamically stable varifold cannot have a tangent cone consisting of more than three half-planes, counting multiplicities. This is because, if there is a tangent cone with two planes, for example, it seems that one can reduce the mass again by a suitable Lipschitz map. But this is rather intricate since being asymptotically close to two planes in measure may still leave lots of complication. One can also ask if there is any classification of tangent flows for Brakke flows obtained in [22]. This is a hard question but quite relevant to regularity theory of general varifolds which has been studied mostly under time-independent situations.

There are numerous existence results which are closely related, even restricting to general Brakke flows. The Brakke flow appears naturally as rigorous singular perturbation limits of the Allen–Cahn equation [18, 36] and the parabolic

Ginzburg–Landau equation [6, 26].[2] Related to the level-set method [9, 13], it is known that the a.e. level-set of viscosity solution of the MCF is a unit density Brakke flow [14]. There is also the elliptic regularization method to obtain Brakke flow, see [19, 32, 43], for example. This method has a nice property that it is relatively simple to construct a solution and that there exists a sequence of smooth (except for a closed singularity with small dimension) approximating MCFs. The class of Brakke flow constructed through the elliptic regularization method is closed under the topology discussed in Sect. 3.3, and one can even construct the Brakke flow with the setting of a flat chain with coefficients in a finite group (see [32]).

There seems to be a zoo of generalized MCFs even among the Brakke flows! The reader may wonder if there could be a definitive subclass of Brakke flow better than any others. Maybe yes for the case of $k = 1$, but I suspect that the answer is no in general. When we think of the Plateau problem of finding area-minimizing surfaces, there are varieties of classes of solutions. There are choices of oriented versus unoriented, fixed topology or not, and we may choose currents, flat chains, closed sets, and others depending on how to approach the problem. If not all perhaps, but one may associate the solution with a stationary varifold in general since the concept of a varifold is perhaps minimally equipped with no extra structures. The definition of Brakke flow is perhaps as minimal as generalized MCFs should all satisfy. Depending on the problems one wants to solve, there should be some number of good subclasses to consider, just like the Plateau problem.

---

[2]Technically speaking, for the limit of the parabolic Ginzburg–Landau equation, the a.e. integrality of the limit varifolds has not been proved.

Chapter 5
Allard Regularity Theory

5.1 Time-Independent Brakke Flows

Before we go into a full-fledged regularity theory for Brakke flows, it is certainly
reasonable to consider a simpler time-independent situation. Suppose that we have
a varifold $V \in \mathbf{V}_k(U)$ which happens to be a time-independent Brakke flow as we
defined in Sect. 2.2. This should mean that the normal velocity v is 0 and that $v = h$
implies $h = 0$, which means that V is stationary. Let us adhere to the definition of
the Brakke flow as in Definition 2.2 and check if this is indeed the case. If V is a
time-independent Brakke flow, the assumptions reduce to:

(i) $V \in \mathbf{IV}_k(U)$,
(ii) V has locally bounded first variation and $\|\delta V\| \ll \|V\|$ in U,
(iii) $h(V, \cdot) \in L^2_{loc}(\|V\|)$ in U,
(iv) for $\phi \in C^1_c(U; \mathbb{R}^+)$, we have

$$0 \le \int_U (\nabla \phi - \phi h(V, \cdot)) \cdot h(V, \cdot) \, \mathrm{d}\|V\|. \tag{5.1}$$

Let us check that $h(V, \cdot) = 0$ follows from above. Since $V \in \mathbf{IV}_k(U)$, we have
$V = \theta | \Gamma$ for a $\mathcal{H}^k$ measurable countably k-rectifiable set Γ and an $\mathcal{H}^k$ measurable
positive integer-valued multiplicity θ defined on Γ. Thus we may think $\|V\| = \theta \mathcal{H}^k \llcorner_\Gamma$. Recall that Γ then has a unique approximate tangent space for $\mathcal{H}^k$ a.e. $x \in \Gamma$. The function θ has the Lebesgue points $\mathcal{H}^k$ a.e. on Γ since Γ is covered by
a countable union of k-dimensional C^1 submanifolds. The same can be said about
$h(V, \cdot)$ and $|h(V, \cdot)|^2$, that is, both $h(V, \cdot)$ and $|h(V, \cdot)|^2$ have the Lebesgue points
$\mathcal{H}^k$ a.e. on Γ. In addition, we have $h(V, \cdot)$ perpendicular to the approximate tangent
space. Fix $x_0 \in \Gamma$ such that all these four conditions hold, which is $\mathcal{H}^k$ a.e. on Γ.
Now we proceed just as we did when we justified the weak formulation of normal
velocity. Namely, given arbitrary $\phi \in C^1_c(\mathbb{R}^n; \mathbb{R}^+)$ and $r > 0$, define $\phi_r(x) :=$

© The Author(s), under exclusive license to Springer Nature Singapore Pte Ltd. 2019
Y. Tonegawa, *Brakke's Mean Curvature Flow*, SpringerBriefs in Mathematics,
https://doi.org/10.1007/978-981-13-7075-5_5

$r^{-k+1}\phi((x - x_0)/r)$. For sufficiently small r, we have $\phi_r \in C^1_c(U; \mathbb{R}^+)$ and we may substitute ϕ_r into (5.1). Suppose that $\mathrm{spt}\,\phi \subset B_R$ and thus $\mathrm{spt}\,\phi_r \subset B_{rR}(x_0)$. We have

$$\lim_{r \to 0+} \int_U \phi_r |h(V, \cdot)|^2 \, \mathrm{d}\|V\| \le \lim_{r \to 0+} r^{-k+1} \sup |\phi| \int_{B_{rR}(x_0)} |h(V, \cdot)|^2 \, \mathrm{d}\|V\| = 0$$

(5.2)

since x_0 is a Lebesgue point of $|h(V, \cdot)|^2$, thus $\int_{B_{rR}(x_0)} |h(V, \cdot)|^2 \, \mathrm{d}\|V\| = O(r^k)$. Due to the existence of approximate tangent space at x_0, we also have after a change of variable that

$$\lim_{r \to 0+} \int_U \nabla \phi_r \cdot h(V, \cdot) \, \mathrm{d}\|V\| = \int_{T_{x_0} \Gamma} \nabla \phi(x) \cdot h(V, x_0) \theta(x_0) \mathrm{d}\mathcal{H}^k(x).$$

(5.3)

Thus we obtain from (5.1) to (5.3) that

$$0 \le \int_{T_{x_0} \Gamma} \nabla \phi(x) \, \mathrm{d}\mathcal{H}^k(x) \cdot h(V, x_0) \theta(x_0).$$

(5.4)

Since $h(V, x_0)$ is perpendicular to $T_{x_0} \Gamma$, (5.4) implies (just like (2.8)) that $h(V, x_0) = 0$. Thus, after all, $h(V, \cdot) = 0$ for $\|V\|$ a.e. on U and V is a stationary integral varifold. Of course, any stationary integral varifold is a time-independent Brakke flow, but we now see that the converse is also true and we state it as follows.

Proposition 5.1 *An integral varifold $V \in \mathbf{IV}_k(U)$ is stationary if and only if it is a time-independent Brakke flow.*

Let us summarize known results on general stationary integral varifolds. For $k = 1$, as we mentioned already, [2] showed that $\mathrm{spt}\,\|V\|$ consists of locally finite line segments with their end points meeting at junctions in a way that they are stationary. The integer multiplicity is constant on each line segment and there are no structural complications. For $k \ge 2$, we know from the Allard regularity theorem [1] that on a dense open subset of $\mathrm{spt}\,\|V\|$, it is a real analytic k-dimensional minimal surface. Stated differently, take any point $x \in \mathrm{spt}\,\|V\|$ and $r > 0$. Then we know that there is a ball $B_{r'}(y) \subset B_r(x)$ in which $\mathrm{spt}\,\|V\| \cap B_{r'}(y)$ is a real analytic k-dimensional minimal surface. On the other hand, the set of points in $\mathrm{spt}\,\|V\|$ which are not part of such regular surfaces may have positive $\mathcal{H}^k$ measure. Such a set may be termed the singular set. Ideally, we would like to say that the singular set has null $\mathcal{H}^k$ measure, or even better, locally finite $\mathcal{H}^{k-1}$ measure, but little is known in general so far.[1] For the rest, we will be more or less focused on the unit density situation, the reason being that even the special static case is not well-understood so far.

---

[1] For some recent progress in this direction, see [31].

5.2   The Allard Regularity Theorem

In this section, we review the Allard regularity theorem. To present the content, I first describe the theorem for a rather special case for the reader to get the idea clearly, and then proceed to describe the full-strength version. So suppose $V \in \mathbf{IV}_k(U)$ is a unit density stationary varifold. The Allard regularity theorem claims, roughly speaking, that whenever V is locally sufficiently close to a unit density k-dimensional affine plane in measure, then spt $\|V\|$ is a regular k-dimensional minimal surface near that point. Since there is no loss of generality to consider the point in question as the origin, we may think this affine plane to be some $S \in \mathbf{G}(n, k)$. It is convenient to define for $r > 0$

$$C(S, r) := \{x \in \mathbb{R}^n : \operatorname{dist}(x, S^{\perp}) < r\},$$

which is a cylinder perpendicular to S. The theorem claims that if $\|V\| \approx \mathcal{H}^k \llcorner_S$ in $C(S, 2)$, then, spt$\|V\| \cap C(S, 1)$ is a regular k-dimensional minimal surface. It is topologically a k-dimensional disk which can be represented as a regular graph over $S \cap C(S, 1)$. This is a very powerful and even surprising result, considering how little we assumed on V in terms of the regularity. A priori, we do not know spt $\|V\|$ is locally a graph, but if it is close to a k-dimensional plane in measure, the theorem claims it is necessarily a graph. Now we describe a more precise statement of the theorem. Without loss of generality after orthogonal rotation, let us assume that

$$S = \mathbb{R}^k \times \{0\} \subset \mathbb{R}^n.$$

Then we have

$$S \cap C(S, r) = B_r^k \times \{0\} \cong B_r^k$$

for $r > 0$. To express "the closeness" to S in $C(S, 2)$, it is convenient to use the following quantity μ defined by:

$$\mu = \left(\int_{C(S,2)} \operatorname{dist}(x, S)^2 \, \mathrm{d}\|V\|(x) \right)^{\frac{1}{2}}. \tag{5.5}$$

Here $\operatorname{dist}(x, S)$ is the distance of x from S and is equal to $|S^{\perp}(x)|$, so we interchangeably use it. In fact, for $x = (x_1, \cdots, x_n)$, we have

$$|S^{\perp}(x)|^2 = \sum_{j=k+1}^{n} (x_j)^2.$$

If $\|V\| = \mathcal{H}^k \llcorner_S$, then $\mu = 0$. If $k = n - 1$ and spt $\|V\|$ is represented as a graph $x_n = f(x_1, \cdots, x_{n-1})$, then since $|S^\perp(x)| = |x_n|$,

$$\mu = \left(\int_{B_2^{n-1}} f^2 \sqrt{1 + |\nabla f|^2} \, d\mathcal{H}^{n-1} \right)^{\frac{1}{2}}. \tag{5.6}$$

We have a similar formula for $k < n - 1$, where we have $x_{k+1} = f_{k+1}(x_1, \cdots, x_k)$, $\cdots, x_n = f_n(x_1, \cdots, x_k)$ and the Jacobian is replaced by the suitable k-dimensional version of the order $1 + O(|\nabla f|^2)$ when $|\nabla f|$ is small. Thus, if $|\nabla f|$ is small, μ behaves like an L^2 norm of the height of the graph relative to S. Since $\mathrm{dist}(x, S)^2$ is a continuous function, if a stationary varifold satisfies $\|V\| \approx \mathcal{H}^k \llcorner_S$, μ will be small. Under suitable assumptions, the Allard regularity theorem asserts that the converse is also true.[2]

Theorem 5.2 *Given $v \in (0, 1)$, there exist constants $\varepsilon \in (0, 1)$ and $\gamma \in (0, 1)$ depending only on v, n, k with the following property. Suppose $V \in \mathbf{IV}_k(C(S, 1))$ is a unit density stationary varifold[3] such that:*

(i) $\|V\|(C(S, 1/2)) \geq v2^{-k}\omega_k$,
(ii) $\|V\|(C(S, 1)) \leq (2 - v)\omega_k$,
(iii) $\mu := \left(\int_{C(S,1)} \mathrm{dist}(x, S)^2 \, d\|V\|(x) \right)^{\frac{1}{2}} \leq \varepsilon$.

Then there exist real analytic functions $f_j : B_\gamma^k \to \mathbb{R}$ $(j = k+1, \cdots, n)$ such that, writing $\hat{x} = (x_1, \cdots, x_k)$,

$$\{(\hat{x}, f_{k+1}(\hat{x}), \cdots, f_n(\hat{x})) : \hat{x} \in B_\gamma^k\} = C(S, \gamma) \cap \mathrm{spt} \|V\| \tag{5.7}$$

and the graph of f is a real analytic minimal surface. Furthermore, given $m \in \mathbb{N}$, there exists a constant c depending only on m, n, k such that

$$\sup_{k+1 \leq j \leq n} \|f_j\|_{C^m(B_\gamma^k)} \leq c \, \mu. \tag{5.8}$$

Besides the smallness of μ, we added two more assumptions (i) and (ii). The assumption (i) is necessary to avoid the case of $C(S, 1/2) \cap \mathrm{spt} \|V\| = \emptyset$. We can replace this condition with other equivalent assumptions as long as we may avoid such a trivial situation, for example, we may instead assume $0 \in \mathrm{spt} \|V\|$ in (i). This empty case cannot be excluded by having only (ii) and (iii). The assumption (ii) is also necessary. It requires that Γ (where $V = |\Gamma|$) has strictly less k-dimensional

---

[2]The assumptions are slightly different from [1, Theorem 8.19] and we take the version of [33, Theorem 23.1] here.

[3]The varifold need not be unit density or integral, but for simplicity and relevance to Brakke flow, we restrict our attention to a unit density varifold.

area (recall $\|V\| = \mathcal{H}^k \llcorner_\Gamma$) than two times that of a k-dimensional disk of radius 1. For example, think of Γ being a union of $S = \mathbb{R}^k \times \{0\}$ and $\tilde{S} = \mathbb{R}^k \times (\delta, 0, \cdots, 0)$, where $\delta \neq 0$ is a small number. So $\Gamma = S \cup \tilde{S}$ is composed of two parallel planes S and $\tilde{S}$, and μ can be made as small as one likes by choosing sufficiently small δ. (i) is satisfied, but not (ii) since $\mathcal{H}^k(\Gamma \cap C(S, 1)) = 2\omega_k$ for any $\delta \neq 0$. Of course, such Γ cannot be represented as a single-valued graph over S. We can also consider a catenoid (which is a minimal surface) with very small hole, which can be made arbitrarily close to two parallel planes in measure and which cannot be represented as a graph over S.

The theorem not only gives the regularity of spt $\|V\|$, but also the derivative estimates of any order in terms of constant multiples of μ. The estimate is analogous to that of harmonic functions. Recall that any interior C^m norms of a harmonic function can be bounded by constant multiples of the L^2 norm of the function (see for example [16]), and (5.8) is similar if one considers μ as an L^2 norm of the height of the graph in a generalized sense.

Next we review the full-fledged Allard regularity theorem. Since stationarity means that $h(V, x) = 0$, one can expect that a similar regularity theorem should hold with some perturbation for $h(V, x)$ in a suitable sense.

Theorem 5.3 *Given $v \in (0, 1)$ and $p \in (k, \infty)$, there exist constants $\varepsilon \in (0, 1)$, $\gamma \in (0, 1)$ and $c \in (1, \infty)$ depending only on v, p, n, k with the following properties. Suppose $V \in \mathbf{IV}_k(C(S, 1))$ is a unit density varifold[4] with*

(i) $\|V\|(C(S, 1/2)) \geq v2^{-k}\omega_k$,

(ii) $\|V\|(C(S, 1)) \leq (2 - v)\omega_k$,

(iii) $\|\delta V\| \ll \|V\|$ *in* $C(S, 1)$ *and* $\|h\|_{L^p} := \left(\int_{C(S,1)} |h(V, x)|^p \, d\|V\|(x) \right)^{\frac{1}{p}} \leq \varepsilon$,

(iv) $\mu := \left(\int_{C(S,1)} \text{dist}(x, S)^2 \, d\|V\|(x) \right)^{\frac{1}{2}} \leq \varepsilon$.

Define

$$\alpha := 1 - \frac{k}{p} > 0.$$

Then there exist $C^{1,\alpha}$ functions $f_j : B_\gamma^k \to \mathbb{R}$ ($j = k + 1, \ldots, n$) representing spt$\|V\|$ *as in (5.7) and*

$$\sup_{k+1 \leq j \leq n} \|f_j\|_{C^{1,\alpha}(B_\gamma^k)} \leq c(\mu + \|h\|_{L^p}). \tag{5.9}$$

The case of $h = 0$ reduces to the previous theorem, except that the claim for the estimate is only for the $C^{1,\alpha}$ norm for all $\alpha < 1$. It is a standard fact for elliptic

---

[4]A similar remark applies about the unit density assumption as in Theorem 5.2.

regularity theory to go from $C^{1,\alpha}$ to real analyticity for minimal surface equations with estimates, so the essential part of the proof for both theorems is the fact that spt$\|V\|$ is a graph with $C^{1,\alpha}$ regularity. If h happens to be more regular, say, $h \in C^\beta$ for some $\beta \in (0, 1)$ on spt$\|V\|$, then we have $f_j \in C^{2,\beta}$ with the corresponding estimates for the $C^{2,\beta}$ norm of f_j in terms of μ and $\|h\|_{C^\beta}$.

A rough interpretation of the estimate in Theorem 5.3 is that, if we already know that spt$\|V\|$ is represented as a graph, we may think of h as a Laplacian of the graph function, since mean curvature is the sum of principal curvatures. If the L^p norm of the Laplacian is bounded, then the standard L^p theory tells us that the graph is in $W^{2,p}$ with an estimate in terms of the L^2 norm of the graph and L^p norm of the Laplacian. If $p > k$, then by the standard Sobolev imbedding theorem, we have an estimate of the $C^{1,\alpha}$ norm in terms of the $W^{2,p}$ norm, with α coming from the Sobolev imbedding. Thus the estimate (5.9) is reasonable in this sense, even though it certainly does not prove the theorem.

For stationary varifolds, the following monotonicity formula is fundamental. A similar formula holds for varifolds with $h(V, \cdot) \in L^p_{\text{loc}}(\|V\|)$, $p > k$ (see [1, 33]).

Proposition 5.4 *Suppose that $V \in \mathbf{V}_k(U)$ is stationary. Then for any fixed $x \in U$, $0 < r_1 < r_2 < \text{dist}(x, \partial U)$, we have*

$$r_1^{-k}\|V\|(B_{r_1}(x)) \le r_2^{-k}\|V\|(B_{r_2}(x)).$$

The easy consequence is the existence of density at each point, namely,

$$\text{for each point } x \in U, \ \theta^k(\|V\|, x) = \lim_{r \to 0} \frac{1}{\omega_k r^k}\|V\|(B_r(x)) \text{ exists.}$$

We also have:

Proposition 5.5 *Suppose that V is a unit density stationary varifold $V = |\Gamma|$ in U. Then*

$$\mathcal{H}^k(\text{spt}\|V\| \Delta \Gamma) = 0$$

and for all $x \in \text{spt}\|V\|$ and $B_r(x) \subset U$,

$$\|V\|(B_r(x)) \ge \omega_k r^k. \tag{5.10}$$

Proof $\mathcal{H}^k(\Gamma \setminus \text{spt}\|V\|) = 0$ is obvious since spt$\|V\|$ is closed by definition. By the rectifiability of Γ, we have $\theta^k(\|V\|, x) = 1$ for $\mathcal{H}^k$ a.e. on Γ. At such a point, by the monotonicity formula, we have

$$\frac{1}{\omega_k r^k}\|V\|(B_r(x)) \ge \lim_{s \to 0} \frac{1}{\omega_k s^k}\|V\|(B_s(x)) = \theta^k(\|V\|, x) = 1$$

as long as $B_r(x) \subset U$. For $\mathcal{H}^k(\mathrm{spt}\|V\| \setminus \Gamma) = 0$, fix arbitrary $x \in \mathrm{spt}\|V\|$. Then by definition, for all small $r > 0$, $\|V\|(B_r(x)) = \mathcal{H}^k(\Gamma \cap B_r(x)) > 0$, thus there is a sequence $\{x_j\}_{j=1}^{\infty}$ such that $\lim_{j \to \infty} x_j = x$ and $\theta^k(\|V\|, x_j) = 1$ for all j. By the monotonicity formula, we have $\|V\|(B_r(x_j)) \geq \omega_k r^k$ for $B_r(x_j) \subset U$, and it follows that $\|V\|(B_r(x)) \geq \omega_k r^k$ for all $B_r(x) \subset U$ as well, proving (5.10). This also shows that $\theta^k(\|V\|, x) \geq 1$ holds whenever $x \in \mathrm{spt}\|V\|$. By a general measure theoretic result (see [12]), for $\mathcal{H}^k$ a.e. $x \in U \setminus \Gamma$, we have

$$\theta^k(\mathcal{H}^k \llcorner_{\Gamma}, x) = \theta^k(\|V\|, x) = 0.$$

Hence we have $\mathcal{H}^k(\mathrm{spt}\|V\| \setminus \Gamma) = 0$. $\qquad\square$

In particular, for a stationary unit density varifold $V = |\Gamma|$, we can always assume without loss of generality that $\Gamma = \mathrm{spt}\|V\|$ and that Γ is a closed set. In addition, since Γ is countably k-rectifiable, Γ has the approximate tangent space $\mathcal{H}^k$ a.e. for $x \in \Gamma$. This means that if one sufficiently magnifies Γ centered at such a point, we may apply Theorem 5.2. More precisely, suppose that Γ has the approximate tangent space $S \in \mathbf{G}(n, k)$ at 0. This means that $\lim_{r \to 0+}(\tau_r)_\sharp V = |S|$. Then, for any $v \in (0, 1)$ fixed, say for $v = 1/2$, the conditions (i)–(iii) of Theorem 5.2 for $(\tau_r)_\sharp V$ are satisfied for sufficiently small r, and there exists a neighborhood of 0 such that $\mathrm{spt}\|V\|$ is real analytic and minimal. Thus we have:

Theorem 5.6 *Suppose $V \in \mathbf{V}_k(U)$ is a unit density stationary varifold. Then there exists a closed set $C \subset \mathrm{spt}\|V\|$ such that $\mathcal{H}^k(C) = 0$ and $\mathrm{spt}\|V\| \setminus C$ is a real analytic embedded k-dimensional minimal surface.*

5.3 A Glimpse at the Proof of the Allard Theorem

I would like to mention a few key elements of the proof for the Allard theorem to compare to the regularity theorem for the Brakke flow. First, it is well-known that the key estimates for various regularity theories in PDEs are often in the form of "Caccioppoli estimates" obtained from the equation, where higher–order derivatives are bounded in terms of lower–order terms. Let us see some analogy in a simple situation of

$$\Delta f = 0 \text{ on } U.$$

Let $\phi \in C_c^1(U)$ be arbitrary. Multiplying the equation with $f\phi^2$ and by integration by parts, we have

$$\int_U |\nabla f|^2 \phi^2 = -\int_U 2f\phi\nabla\phi \cdot \nabla f.$$

By the Cauchy–Schwarz inequality,

$$\int_U |\nabla f|^2 \phi^2 \le \frac{1}{2} \int_U |\nabla f|^2 \phi^2 + 2 \int_U |\nabla \phi|^2 f^2$$

and

$$\int_U |\nabla f|^2 \phi^2 \le 4 \int_U |\nabla \phi|^2 f^2, \tag{5.11}$$

which gives control of the L^2 norm of the gradient ∇f in terms of the L^2 norm of f. For a stationary varifold V in $U \subset \mathbb{R}^n$, the analogous computation can be seen as follows. In place of $\Delta f = 0$, we have the equation $\delta V(g) = 0$ for any $g \in C_c^1(U; \mathbb{R}^n)$. Let $\phi \in C_c^1(U; \mathbb{R}^+)$ be arbitrary. Let g be a vector field defined by

$$g(x) = S^\perp(x)\phi(x)^2.$$

Here $S \in \mathbf{G}(n, k)$ is fixed. Note the analogy between g and $f\phi^2$ which we saw before. One can think that $S^\perp(x)$ is a substitute for f since $S^\perp(x)$ evaluated by the measure $\|V\|$ should behave more or less like the vector-valued height of a graph over the plane S. To compute

$$\delta V(g) = \int_{\mathbf{G}_k(U)} W \cdot \nabla g \, dV(x, W),$$

we have for any $W \in \mathbf{G}(n, k)$

$$W \cdot \nabla(S^\perp(x)\phi(x)^2) = 2\phi(x) W \cdot (S^\perp(x) \otimes \nabla\phi(x)) + \phi(x)^2 W \cdot S^\perp$$

$$= 2\phi(x) \sum_{i,j=1}^n W_{ij}(S^\perp(x))_i \nabla_{x_j}\phi(x) + \phi(x)^2 W \cdot S^\perp$$

$$= 2\phi(x) W(S^\perp(x)) \cdot \nabla\phi(x) + \phi(x)^2 W \cdot S^\perp. \tag{5.12}$$

Here we used $\nabla S^\perp(x) = S^\perp$ in the first line and also the explicit matrix indices in the second line. For $W, S \in \mathbf{G}(n, k)$,

$$|W(S^\perp(x))| = |(W \circ (I - S))(S^\perp(x))|$$

$$= |(W \circ (W - S))(S^\perp(x))| \tag{5.13}$$

$$\le \|W - S\| |S^\perp(x)|$$

since $W = W \circ W$ and $\|W\| \le 1$. We also have

$$\|W - S\|^2 \le 2 W \cdot S^\perp. \tag{5.14}$$

This is because, if we choose $v \in \mathbb{R}^n$ with $|v| = 1$ such that $\|W - S\| = (W - S)(v)$, we have

$$\|W - S\|^2 = |(W - S)(v)|^2 = v^\top \cdot ((W - S) \circ (W - S))(v) \le (W - S) \cdot (W - S)$$

since these matrices are symmetric and $(W - S) \circ (W - S)$ is non-negative. On the other hand, since $W \cdot W = k$ for any $W \in \mathbf{G}(n, k)$, we have

$$(W - S) \cdot (W - S) = 2k - 2W \cdot S = 2W \cdot I - 2W \cdot S = 2W \cdot S^\perp.$$

These show (5.14). Thus combining (5.12), (5.13) and (5.14), we obtain

$$\frac{1}{2}\|W - S\|^2 \phi(x)^2 \le W \cdot \nabla(S^\perp(x)\phi(x)^2)$$

$$+ 2\phi(x)\|W - S\|\|S^\perp(x)\||\nabla\phi(x)|. \tag{5.15}$$

Then integrating (5.15) with respect to $dV(x, W)$ and using

$$\delta V(S^\perp(x)\phi(x)^2) = 0,$$

we have

$$\frac{1}{2}\int_{\mathbf{G}_k(U)} \|W - S\|^2\phi(x)^2 \, dV(x, W)$$

$$\le 2 \int_{\mathbf{G}_k(U)} \phi(x)\|W - S\|\|S^\perp(x)\||\nabla\phi(x)| \, dV(x, W).$$

By the Cauchy–Schwarz inequality, we obtain

$$\int_{\mathbf{G}_k(U)} \|W - S\|^2\phi(x)^2 \, dV(x, W) \le 16 \int_U |\nabla\phi(x)|^2|S^\perp(x)|^2 \, d\|V\|(x). \tag{5.16}$$

When $V = |\Gamma|$, (5.16) is

$$\int_\Gamma \|\mathbf{T}_x\Gamma - S\|^2\phi(x)^2 \, d\mathcal{H}^k(x) \le 16 \int_\Gamma |\nabla\phi(x)|^2 \mathrm{dist}(x, S)^2 \, d\mathcal{H}^k(x). \tag{5.17}$$

The inequality (5.17) is an analogue of (5.11) since $\|\mathbf{T}_x\Gamma - S\|$ behaves like the modulus of the gradient of a graph. In fact, consider the case of $k = n - 1$ and when Γ is represented as a graph $x_n = f(x_1, \ldots, x_{n-1})$. Assume that S is $\mathbb{R}^{n-1} \times \{0\}$. Then it is a good exercise to check that $\|\mathbf{T}_x\Gamma - S\| = \dfrac{|\nabla f|}{\sqrt{1+|\nabla f|^2}}$. The left-hand side is generally called "tilt-excess" in geometric measure theory context, and it measures the deviation of $T_x \Gamma$ from a fixed k-dimensional plane S. The analogy the reader should have in mind is that this quantity is like $\int |\nabla f|^2$ for the graph, even though it is not precisely so due to the nonlinearity. In any case, here the important point is that $\delta V = 0$ gives a certain "Caccioppoli inequality" (5.17).

The second ingredient is the approximation of Γ by Lipschitz functions. Suppose that $V = |\Gamma|$ is a unit density stationary varifold satisfying conditions (i) and (ii) in Theorem 5.2. The claim is that there exists a constant c depending only on k, n, ν such that, if $\mu = (\int_{C(S,1)} \text{dist}(x, S)^2 \, d\|V\|)^{1/2}$ is sufficiently small, there exists a Lipschitz function $\tilde{f} : B_{1/2}^k \to \mathbb{R}^{n-k}$ with Lipschitz constant less than 1 with the property that

$$\mathcal{H}^k(C(S, 1/2) \cap (\Gamma \triangle \text{graph } \tilde{f})) \le c \int_{C(S,3/4) \cap \Gamma} \|T_x \Gamma - S\|^2 \, d\mathcal{H}^k(x). \quad (5.18)$$

The left-hand side represents the error of approximation of Γ by the graph of $\tilde{f}$ and the right-hand side is controlled by $c\mu^2$ due to (5.17). A rather important aspect of this approximation is that this error estimate also gives a $W^{1,2}$ estimate of $\tilde{f}$:

$$\int_{B_{1/2}^k} |\tilde{f}|^2 + |\nabla \tilde{f}|^2 \le c\mu^2. \quad (5.19)$$

The reason for this to be true is that, whenever the graph of $\tilde{f}$ coincides with Γ, $|\tilde{f}(x)| = \text{dist}(\tilde{f}(x), S)$ and $|\nabla \tilde{f}(x)| \le c(n, k)\|T_{\tilde{f}(x)} \Gamma - S\|$, for $\mathcal{H}^k$ a.e. At the places where they do not coincide, we simply use the fact that $|\tilde{f}|$ and $|\nabla \tilde{f}|$ are uniformly bounded since $\tilde{f}$ is Lipschitz function. Since the measure of such a set is bounded by $c\mu^2$, we obtain (5.19).

Finally, for the proof of Theorem 5.2, the crucial ingredient is the following decay estimate.

Proposition 5.7 *Corresponding to $\nu \in (0, 1)$, $\alpha \in (0, 1)$, $n, k \in \mathbb{N}$, there exist $\varepsilon \in (0, 1)$, $c \in (1, \infty)$ and $\lambda \in (0, 1/2)$ with the following property. Suppose $S, W \in \mathbf{G}(n, k)$, $A \in \mathbf{A}(n, k)$ is parallel to S, and V is a unit density stationary varifold V in $C(W, 1)$. Assume that they satisfy:*

(i) $\|V\|(B_{1/2}) \ge \nu 2^{-k} \omega_k$,
(ii) $\|V\|(C(W, 1)) \le (2 - \nu)\omega_k$,
(iii) $\mu := \left(\int_{C(W,1)} \text{dist}(x, A)^2 \, d\|V\|(x) \right)^{\frac{1}{2}} \le \varepsilon$,
(iv) $\|W - S\| \le \varepsilon$.

Then there exist $\tilde{W} \in \mathbf{G}(n, k)$ and $\tilde{A} \in \mathbf{A}(n, k)$ which is parallel to $\tilde{W}$ such that:

(a) $\|\tilde{W} - S\| \le c\mu$,
(b)

$$\left(\lambda^{-(k+2)} \int_{C(W,\lambda)} \operatorname{dist}(x, \tilde{A})^2 \, d\|V\|(x)\right)^{\frac{1}{2}}$$

$$\leq \lambda^{\alpha} \left(\int_{C(W,1)} \operatorname{dist}(x, A)^2 \, d\|V\|(x)\right)^{\frac{1}{2}}.$$

The claim is that there exists a better approximating $\tilde{A}$ with respect to which the normalized L^2 norm of distance function is strictly smaller in the cylinder of radius λ by the factor of λ^{α}. The exponent $k + 2$ of the normalization is correct since the distance function scales like r^2 and the measure scales like r^k under dilations. Also the difference of tilt between $\tilde{A}$ and A (which is $\|\tilde{W} - S\|$) is estimated in terms of μ. The conclusion (b) lets us repeat the use of this proposition again after change of variable $x \to \tilde{x} = x/\lambda$, $V \to \tilde{V} = (\tau_\lambda)_\sharp V$ and replacing A by $\tilde{A}$, because the corresponding μ with respect to $\tilde{A}$ is again less than ε. One has to make sure that the other conditions (i), (ii) and (iv) are satisfied for this rescaled varifold, but that can be done. The iterative use shows that the tangent space of $\operatorname{spt}\|V\|$ is determined at each point, with Hölder estimates on the variation of tangent spaces. This is the key estimate from which $\operatorname{spt}\|V\|$ being a $C^{1,\alpha}$ graph over W with $C^{1,\alpha}$ estimate in terms of μ follows from a rather standard iteration argument.

The proof of Proposition 5.7 is by contradiction. Fix $\nu \in (0, 1)$. Assume that we have $S \in \mathbf{G}(n, k)$ and for each $i \in \mathbb{N}$, assume that we have $A_i \in \mathbf{A}(n, k)$, which is parallel to S, $W_i \in \mathbf{G}(n, k)$ and unit density stationary varifolds $V_i = |\Gamma_i|$ in $C(W_i, 1)$ such that:

 (i) $\|V_i\|(B_{1/2}) \geq \nu 2^{-k}\omega_k$,
 (ii) $\|V_i\|(C(W_i, 1)) \leq (2 - \nu)\omega_k$,
 (iii) $\mu_i := \left(\int_{C(W_i,1)} \operatorname{dist}(x, A_i)^2 \, d\|V_i\|(x)\right)^{\frac{1}{2}} \to 0$ as $i \to \infty$,
 (iv) $\lim_{i \to \infty} \|W_i - S\| = 0$,

and for any $\tilde{W} \in \mathbf{G}(n, k)$ with $\|\tilde{W} - S\| \leq c\mu_i$ and any $\tilde{A} \in \mathbf{A}(n, k)$ parallel to $\tilde{W}$, we have

$$\left(\lambda^{-(k+2)} \int_{C(W_i,\lambda)} \operatorname{dist}(x, \tilde{A})^2 \, d\|V_i\|(x)\right)^{\frac{1}{2}} > \lambda^{\alpha} \mu_i. \tag{5.20}$$

Here, λ is a constant depending only on n, k and α, and c is a constant depending only on n and k. These constants will be determined in the following. If we obtain a contradiction, then that proves the claim of Proposition 5.7. Under this setting, note that for any fixed $r \in (0, 1)$, we have

$$\lim_{i \to \infty} \sup_{x \in \operatorname{spt}\|V_i\| \cap C(S,r)} \operatorname{dist}(x, A_i) = 0, \tag{5.21}$$

that is, the support of $\|V_i\|$ approaches to A_i locally uniformly in $C(S, 1)$. This follows from (5.10): if there are points $x_i \in \operatorname{spt}\|V_i\| \cap C(S, r)$ with $\operatorname{dist}(x_i, A_i) \geq \varepsilon$

for infinitely many i, where we may assume $\varepsilon < (1-r)/2$, then $\|V_i\|(B_{\varepsilon/2}(x_i)) \geq \omega_k(\varepsilon/2)^k$ by (5.10). Then we have

$$\int_{C(W_i,1)} \text{dist}\,(x, A_i)^2 \, d\|V_i\|(x) \geq (\varepsilon/2)^2 \|V_i\|(B_{\varepsilon/2}(x_i)) \geq \varepsilon^{k+2} c(k)$$

for infinitely many i and this contradicts $\mu_i \to 0$. In view of (i), this means that A_i cannot be far away from $B_{1/2}$ for all large i, since otherwise we would have $\|V_i\|(B_{1/2}) = 0$. Since A_i is parallel to S and not far away from the origin, and W_i approaches to S, given $r < 1$, we see that $\text{spt}\,\|V_i\| \cap C(S, r) \subset C(W_i, 1)$ for all large i. In particular, we have

$$\left(\int_{C(S,r)} \text{dist}\,(x, A_i)^2 \, d\|V_i\|(x)\right)^{\frac{1}{2}} \leq \mu_i.$$

Since A_i is parallel to S, by parallel translation, we may assume that $A_i = S$ in the following. Next, as stated before, for all sufficiently small μ_i, there exists a Lipschitz function $\tilde{f}_i$ satisfying (5.19) and (5.18) with respect to Γ_i and μ_i. We define

$$F_i(x) := \frac{\tilde{f}_i(x)}{\mu_i}$$

and note that we have from (5.19)

$$\int_{B_{1/2}^k} |F_i(x)|^2 + |\nabla F_i(x)|^2 \leq c(n, k, \nu), \tag{5.22}$$

which is independent of i. By the standard compactness theorem for Sobolev space, there exists a subsequence (denoted by the same index) and a limit function F such that F_i converges weakly to F in $W^{1,2}(B_{1/2}^k)$ and strongly in $L^2(B_{1/2}^k)$. The important claim is that the limit function F is harmonic. Consider the case of $k = n - 1$ since it is easier to see. All we need to prove is that we have

$$\int_{B_{1/2}^{n-1}} \nabla F \cdot \nabla \phi = 0$$

for any $\phi \in C_c^1(B_{1/2}^{n-1})$ to conclude that F is harmonic. For this, we use $\delta V_i(g) = 0$ with vector field $g = (0, \ldots, 0, \phi(S(x)))$. Note that this g does not have a compact support, but it is fine since $\text{spt}\,\|V_i\|$ is within a small neighborhood of S due to (5.21). Then we have

$$0 = \delta V_i(g) = \int_{\Gamma_i} T_x \Gamma_i \cdot \nabla g \, d\mathcal{H}^{n-1}.$$

Due to (5.18), replacing Γ by the graph of $\tilde{f}_i$ results in some error of at most $c\mu_i^2$. Hence we have

$$\left| \int_{B_{1/2}^{n-1}} T_{\tilde{f}_i(x)} \operatorname{graph} \tilde{f}_i \cdot \nabla g \right| \le c\mu_i^2. \tag{5.23}$$

Since the orthogonal projection to the tangent space to graph $\tilde{f}_i$ is given by $I - \nu \otimes \nu$ with $\nu = (-\nabla \tilde{f}_i, 1)/\sqrt{1 + |\nabla \tilde{f}_i|^2}$, one can check that the integrand of (5.23) is $\nabla \tilde{f}_i \cdot \nabla \phi/(1 + |\nabla \tilde{f}_i|^2)$. Since the error of replacing $1 + |\nabla \tilde{f}_i|^2$ by 1 results in an error of $c\mu_i^2$ by (5.19), we see that

$$(\mu_i)^{-1} \left| \int_{B_{1/2}^{n-1}} \nabla \tilde{f}_i \cdot \nabla \phi \right| \to 0$$

as $i \to \infty$. Since $(\mu_i)^{-1} \nabla \tilde{f}_i = \nabla F_i$ converges weakly to ∇F, this proves the claim that F is a harmonic function. The computation is more or less the same for general $k < n - 1$. Now the idea to get to a contradiction is that we choose $\tilde{A}$, which is given by the linear approximation of F at the origin. For harmonic function F, the well-known property is that all the derivatives of F in the interior are bounded in terms of the L^1 norm of F, so in particular

$$\sup_{x \in B_{1/4}^k} |\nabla^2 F(x)| \le c(k, n) \int_{B_{1/2}^k} |F|.$$

We let $\tilde{A}_i$ be the k-dimensional affine plane given as a graph of

$$\mu_i F(0) + \mu_i x \cdot \nabla F(0).$$

We have the same inequality (5.22) for F and within B_λ^k ($0 < \lambda < 1/4$), thus

$$\sup_{B_\lambda^k} |F(x) - F(0) - x \cdot \nabla F(0)| \le \lambda^2 \sup_{B_\lambda^k} |\nabla^2 F| \le c(n, k, \nu)\lambda^2 \tag{5.24}$$

holds. For

$$\lambda^{-(k+2)} \int_{C(S,\lambda)} \operatorname{dist}(x, \tilde{A}_i)^2 \, \mathrm{d}\|V_i\|(x)$$

$$= \lambda^{-(k+2)} \int_{C(S,\lambda) \cap \Gamma_i} \operatorname{dist}(x, \tilde{A}_i)^2 \, \mathrm{d}\mathcal{H}^k(x),$$

replacing Γ_i by the graph of $\tilde{f}_i$ results in an error of $c\mu_i^2$ which is multiplied by dist $(\mathrm{spt}\,\|V_i\|, \tilde{A}_i)^2$. By (5.21) and $\tilde{A}_i \to S$, we have dist$(\mathrm{spt}\,\|V_i\|, \tilde{A}_i) \to 0$, thus this replacement error goes to 0 even after dividing by μ_i^2 as $i \to \infty$. Thus, we have

$$\limsup_{i\to\infty} (\mu_i)^{-2}\lambda^{-(k+2)} \int_{C(S,\lambda)} \mathrm{dist}\,(x, \tilde{A}_i)^2 \,\mathrm{d}\|V_i\|$$

$$\leq \limsup_{i\to\infty} (\mu_i)^{-2}\lambda^{-(k+2)} \int_{B_\lambda^k} |\tilde{f}_i(x) - \mu_i(F(0) + x \cdot \nabla F_i(0))|^2$$

$$\leq 2\lim_{i\to\infty} \lambda^{-(k+2)} \int_{B_\lambda^k} |F_i(x) - F(x)|^2$$

$$+ 2\lambda^{-(k+2)} \int_{B_\lambda^k} |F(x) - F(0) - x \cdot \nabla F(0)|^2.$$

By the strong L^2 convergence, the first term is 0. By (5.24), the second term is bounded by $c(n, k, v)\lambda^2$. Then we may choose λ small depending only on n, k, v, α such that $c(n, k, v)\lambda^2 < \lambda^{2\alpha}/4$, proving

$$\limsup_{i\to\infty} (\mu_i)^{-1}\left(\lambda^{-(k+2)} \int_{C(S,\lambda)} \mathrm{dist}\,(x, \tilde{A}_i)^2 \,\mathrm{d}\|V_i\|\right)^{\frac{1}{2}} \leq \lambda^\alpha/2.$$

Here, the domain of integration is not $C(W_i, \lambda)$. However, this is not a problem since the tilt difference between W_i and S gets smaller as $i \to \infty$, so $C(W_i, \lambda) \subset C(S, \lambda + \epsilon)$ for any $\epsilon > 0$ and for all large i and we can equally prove

$$\limsup_{i\to\infty} (\mu_i)^{-1}\left(\lambda^{-(k+2)} \int_{C(W_i,\lambda)} \mathrm{dist}\,(x, \tilde{A}_i)^2 \,\mathrm{d}\|V_i\|\right)^{\frac{1}{2}} \leq \lambda^\alpha/2.$$

Note also that for $\tilde{W}_i \in \mathbf{G}(n, k)$ which is parallel to $\tilde{A}_i$,

$$\|S - \tilde{W}_i\| \leq c(n, k)\mu_i|\nabla F(0)| \leq c(k, n)\mu_i$$

due to the way $\tilde{A}_i$ is defined. This leads to a contradiction to (5.20) with $\tilde{W} = \tilde{W}_i$ and $\tilde{A} = \tilde{A}_i$ for all sufficiently large i, concluding the proof of Proposition 5.7.

Chapter 6
Regularity Theory for the Brakke Flow

6.1 Main Regularity Theorems

The regularity theorem for a Brakke flow is similar to the Allard regularity theorem. Roughly speaking, the claim is that the support is smooth in the interior whenever it is sufficiently close in measure to a k-dimensional plane in a space–time neighborhood.

Let us first discuss what should be the corresponding assumptions for the Brakke flow in view of (i)–(iii) of Theorem 5.2. Assume that we have a unit density Brakke flow $\{V_t\}_{t\in(-T,T)}$ in $C(S, 1)$ for unspecified $T > 0$ for the moment. A good guess should be that, if we assume V_t is close to S, then spt $\|V_t\|$ is smooth in some $C(S, \gamma)$ near $t = 0$, with a suitable set of assumptions. Assumption (i) of Theorem 5.2 prevents the case that $\|V\| = 0$. For the Brakke flow, obviously, we want to put something similar to prevent $\|V_t\| = 0$ as part of the assumption. One particular aspect of the Brakke flow is that, even if $V_t \neq 0$ at the "beginning" of the time interval $(-T, T)$, it is possible that $V_t = 0$ soon after that, since we can always set $V_t = 0$ for $t \geq t_0$ for arbitrary $t_0 \in (-T, T)$. So it is natural to assume some type of assumption "$V_t \neq 0$" near the end of the time interval if we want to make sure that $V_t \neq 0$ around $t = 0$. Assumption (ii) of Theorem 5.2 prevents the case of two parallel k-dimensional planes. The situation is the same for the Brakke flow since two parallel k-dimensional planes with no motion give a Brakke flow as well, but it is not possible to represent it as a single-valued graph. It is reasonable to assume that $\|V_t\|(C(S, 1))$ is strictly smaller than $2\omega_k$ near the "beginning" of $(-T, T)$ since the measure is supposed to be more or less decreasing in time. There can be a number of ways to express these two assumptions (i) and (ii), but it turned out that it is technically convenient to use the following function. Let $\phi \in C^\infty(\mathbb{R}^+)$

© The Author(s), under exclusive license to Springer Nature Singapore Pte Ltd. 2019
Y. Tonegawa, *Brakke's Mean Curvature Flow*, SpringerBriefs in Mathematics,
https://doi.org/10.1007/978-981-13-7075-5_6

be a smooth approximation of the characteristic function of $[0, 3/4]$ so that $\phi = 1$ on $[0, 2/3]$, $\phi = 0$ on $[5/6, \infty)$, and $0 \le \phi \le 1$ on $\mathbb{R}^+$. For $S \in \mathbf{G}(n, k)$ and $x \in \mathbb{R}^n$, define

$$\phi_S(x) := \phi(|S(x)|).$$

One can think that ϕ_S is a smooth approximation of the characteristic function of cylinder $C(S, 3/4)$ which is radial in the S direction and constant in the $S^\perp$ direction. In addition, we define

$$\mathbf{c} := \int_S \phi_S(x)\, d\mathcal{H}^k(x),$$

which is a value approximately equal to $\omega_k (3/4)^k$. We use ϕ_S to express the analogues of assumptions (i) and (ii) as $\|V_t\|(\phi_S) \ge \nu\mathbf{c}$ and $\|V_t\|(\phi_S) \le (2 - \nu)\mathbf{c}$, respectively, where the first one should be satisfied at the end of the interval and the second one near the beginning. As for the assumption (iii) of Theorem 5.2, it turned out that there are actually a few possibilities, but a simple analogue should be that we ask

$$\mu := \left(\int_{-T}^{T} \int_{C(S,1)} \text{dist}\,(x, S)^2 \, d\|V_t\| dt \right)^{\frac{1}{2}}$$

be small. Note that if $\mu = 0$, then $\int_{C(S,1)} \text{dist}\,(x, S)^2 \, d\|V_t\| = 0$ for a.e. t and we expect that spt $\|V_t\| = S$. With these comments in mind, the following is the first version of the regularity theorem for the Brakke flow.

Theorem 6.1 *Corresponding to $E_1 \in [1, \infty)$ and $\nu \in (0, 1)$, there exist $\varepsilon \in (0, 1)$, $\gamma \in (0, 1)$ and $T \in (2, \infty)$ with the following properties. Suppose that $\{V_t\}_{t \in (-T, T)}$ is a unit density Brakke flow in $C(S, 1)$ such that:*

(i) *for some $t_1 \in [-T + 1/2, -T + 1]$, we have $\|V_{t_1}\|(\phi_S) \le (2 - \nu)\mathbf{c}$,*
(ii) *for some $t_2 \in [T - 1, T - 1/2]$, we have $\|V_{t_2}\|(\phi_S) \ge \nu\mathbf{c}$,*
(iii) *$\mu \le \varepsilon$,*
(iv) *for all $B_r(x) \subset C(S, 1)$ and $t \in (-T, T)$, we have $\|V_t\|(B_r(x)) \le \omega_k r^k E_1$.*

Then there exist C^∞ functions $f_j : B_\gamma^k \times (-1, 1) \to \mathbb{R}\, (j = k+1, \ldots, n)$ such that

$$\{(\hat{x}, f_{k+1}(\hat{x}, t), \ldots, f_n(\hat{x}, t)) : \hat{x} \in B_\gamma^k\} = C(S, \gamma) \cap \text{spt}\,\|V_t\|$$

for all $t \in (-1, 1)$ and the graph is a smooth MCF. Moreover, given $m \in \mathbb{N}$, there exists a constant c depending only on m, n, k, ν such that

$$\sup_{k+1 \le j \le n} \|f_j\|_{C^m(B_\gamma^k \times (-1,1))} \le c\mu,$$

where C^m is the norm in the parabolic sense which counts the order of each t-derivative twice.

Because of Proposition 3.5, we typically have a bound (iv) locally in the domain. If we choose $\nu \in (0, 1)$ closer to 1, it is a more restrictive assumption, but we can have a small ν closer to 0 if we wish. Depending on the chosen E_1 and ν, the claim is that we have possibly small ε and γ and possibly a large T with the properties stated in the theorem. Compared to the Allard regularity theorem, the distinctive difference here is that we have the constant T. From the proof, T may be thought of as a "waiting time" for the measure $\|V_t\|(\phi_S)$ to go down from $(2 - \nu)\mathbf{c}$ to a value closer to a constant multiple of μ, in fact.

Before going further, let us see one application.

Proposition 6.2 *Suppose that a unit density Brakke flow has a tangent flow which is a time-independent unit density k-dimensional plane at a point (x_0, t_0). Then there exists a space–time neighborhood of (x_0, t_0) in which the support of the Brakke flow is a C^∞ MCF.*

This gives the local equivalence between the existence of such tangent flow and the smoothness of the Brakke flow. Let us check the claim.

Proof Assume after a change of variables that $(x_0, t_0) = (0, 0)$ and a Brakke flow V_t has a stated tangent flow $|S|$ with $S \in \mathbf{G}(n, k)$ at $(0, 0)$. We have a local bound of (iv) so that we have some E_1, and let $\nu = 1/2$ be fixed. Then Theorem 6.1 gives ε, γ, T corresponding to these E_1, ν. The fact that $|S|$ is a tangent flow means that there exists a positive sequence $\{r_i\}_{i \in \mathbb{N}}$ such that $\lim_{i \to \infty} r_i = 0$ and $V_t^i = (\tau_{r_i})_\sharp V_{r_i^2 t}$ converges to $|S|$ in the way described in Theorem 3.7. We claim that, given $\epsilon > 0$, we have the following for all sufficiently large i:

$$A_{\epsilon, i, t} := \{x \in C(S, 1) : \epsilon \leq \text{dist}(x, S) \leq 2\} \cap \text{spt}\|V_t^i\| = \emptyset$$

for all $t \in [-T, T]$. This is because of Proposition 3.6, namely, if there is any measure, there must be some definite amount in the space–time neighborhood. To prove it, for a contradiction, suppose that we have a subsequence of index (denoted by the same notation) such that $y_i \in A_{\epsilon, i, t_i}$ for some $t_i \in [-T, T]$. We may assume after choosing a subsequence that $\lim_{i \to \infty} y_i = y$ and $\lim_{i \to \infty} t_i = t$. We have $\epsilon \leq \text{dist}(y, S) \leq 2$ and $t \in [-T, T]$. For $R \in [\epsilon/12, \epsilon/6]$, Proposition 3.6 gives $\|V_{t_i - cR^2}^i\|(B_{3R}(y_i)) \geq \gamma R^k$. Since $t_i \to t$ and $y_i \to y$, this implies that $\|V_{t-cR^2}^i\|(B_{4R}(y)) \geq \gamma R^k/2$ for $R = \epsilon/9$, for example, for all sufficiently large i. On the other hand, $\|V_{t-cR^2}^i\|(B_{4R}(y)) \to 0$ as $i \to \infty$ since $B_{4R}(y)$ is away from S due to $\text{dist}(y, S) \geq \epsilon$ and $R = \epsilon/9$ and $\|V_{t-cR^2}^i\|$ converges to $\||S|\| = \mathcal{H}^k \llcorner_S$. So this is a contradiction. Because of this, we may assume that $\{V_t^i\}_{t \in [-T, T]}$ is a Brakke flow in $\{x \in C(S, 1) : \text{dist}(x, S) < 2\}$ for all sufficiently large i since there is no measure in $\{x \in C(S, 1) : \epsilon \leq \text{dist}(x, S) \leq 2\}$. From the above "uniform convergence", it is clear that

$$\mu^i := \left(\int_{-T}^{T} \int_{C(S,1)} \text{dist}(x, S)^2 \, d\|V_t^i\| \, dt \right)^{\frac{1}{2}}$$

converges to 0 as $i \to \infty$. Thus (iii) is satisfied for all sufficiently large i. Since $\|V_t^i\|(\phi_S) \to \||S|\|(\phi_S) = \int_S \phi_S \, d\mathcal{H}^k = \mathbf{c}$ for all $t \in [-T, T]$, (i) and (ii) with $\nu = 1/2$ are also satisfied for sufficiently large i. Thus, all the conditions of Theorem 6.1 are met, and spt $\|V_t^i\| \cap C(S, \gamma)$ is a smooth MCF for $t \in (-1, 1)$. Going back to the original coordinates, this shows that there exists a space–time neighborhood of (x_0, t_0) where spt $\|V_t\|$ is a C^∞ MCF, proving the claim.   □

In general, we may prove the following "almost everywhere, at almost all times" smoothness.

Theorem 6.3 *For a unit density Brakke flow $\{V_t\}_{t \in [0,T)}$ in $U \subset \mathbb{R}^n$, for a.e. $t \in (0, T)$, there exists a closed set $C_t \subset U$ with the following properties. We have $\mathcal{H}^k(C_t) = 0$ and for any $x \in U \setminus C_t$, there exists a space–time neighborhood $O_{(x,t)} \subset U \times (0, T)$ such that the support of the Brakke flow is either empty in $O_{(x,t)}$ or a C^∞ MCF in $O_{(x,t)}$.*

The proof can be found in [21, 37], as for other theorems in this section. If it is a time-independent Brakke flow, then the result reduces to the same claim as Theorem 5.6. The closed set C_t is a set of singularities which has null k-dimensional measure. Just as in the stationary case, it may or may not be empty. The characterization of "a.e. $t \in (0, T)$" is the time when the measure $\|V_t\|$ is continuous in t, and in fact, the complement of such set of "good times" is countable at most (see [21, Section 9] for the precise characterization).

As in the Allard regularity theorem, we next see the full-strength version. Before we proceed, it is good to have the following analogy in mind. Since the Brakke flow is like the heat equation $\frac{\partial f}{\partial t} = \Delta f$, what we will see next is the analogue of $\frac{\partial f}{\partial t} = \Delta f + u$, where we assume a suitable regularity condition on u such as the space–time integrability natural to the desired result. From the point of view of scaling for this heat equation, since u is like a velocity = "distance/time = x/t", under the parabolic change of variables $x = \lambda \tilde{x}$ and $t = \lambda^2 \tilde{t}$, u should be changed to $u = \lambda^{-1} \tilde{u}$. The integral also changes as

$$\left(\int_0^\infty \left(\int_{\mathbb{R}^k} |u(x, t)|^p \, dx \right)^{\frac{q}{p}} dt \right)^{\frac{1}{q}} = \lambda^{\frac{k}{p} + \frac{2}{q} - 1} \left(\int_0^\infty \left(\int_{\mathbb{R}^k} |\tilde{u}(\tilde{x}, \tilde{t})|^p \, d\tilde{x} \right)^{\frac{q}{p}} d\tilde{t} \right)^{\frac{1}{q}}.$$

For the regularity, as $\lambda \to 0$, the perturbative term $\tilde{u}$ should have a smaller norm so it is natural to ask

$$\zeta := 1 - \frac{k}{p} - \frac{2}{q} > 0. \tag{6.1}$$

With this in mind, the following is a natural generalization of Theorem 6.1.

Theorem 6.4 ([21, Theorem 8.7]) *Suppose $p \in [2, \infty)$ and $q \in (2, \infty)$ satisfies (6.1). Corresponding to $E_1 \in [1, \infty)$, p, q and $\nu \in (0, 1)$, there exist $\varepsilon \in (0, 1)$, $\gamma \in (0, 1)$, $c \in (1, \infty)$ and $T \in (2, \infty)$ with the following properties. Suppose that the unit density $\{V_t\}_{t \in (-T,T)}$ in $C(S, 1)$ satisfies (i)–(iv) of Definition 2.2 and that it* *has a generalized normal velocity* $v(x, t) = h(V_t, x) + u(x, t)^{\perp}$. *Here* $u(x, t)^{\perp}$ *is the normal projection of* $u(x, t)$ *with respect to the approximate tangent space of* V_t *at* x. *Assume that*

(i) *for some* $t_1 \in [-T + 1/2, -T + 1]$, *we have* $\|V_{t_1}\|(\phi_S) \le (2 - \nu)\mathbf{c}$,

(ii) *for some* $t_2 \in [T - 1, T - 1/2]$, *we have* $\|V_{t_2}\|(\phi_S) \ge \nu\mathbf{c}$,

(iii) $\mu := \left(\int_{-T}^{T} \int_{C(S,1)} \text{dist} (x, S)^2 \, \mathrm{d}\|V_t\| \mathrm{d}t \right)^{\frac{1}{2}} \le \varepsilon$,

(iv) *for all* $B_r(x) \subset C(S, 1)$ *and* $t \in (-T, T)$, *we have* $\|V_t\|(B_r(x)) \le \omega_k r^k E_1$,

(v) $\|u\|_{L^{p,q}} := \left(\int_{-T}^{T} \left(\int_{C(S,1)} |u(x, t)|^p \, \mathrm{d}\|V_t\|(x) \right)^{\frac{q}{p}} \mathrm{d}t \right)^{\frac{1}{q}} \le \varepsilon$.

Then there exist $C^{1,\zeta}$ *functions* $f_j : B_{\gamma}^k \times (-1, 1) \to \mathbb{R}$ $(j = k + 1, \dots, n)$ *such that*

$$\{(\hat{x}, f_{k+1}(\hat{x}, t), \dots, f_n(\hat{x}, t)) : \hat{x} \in B_{\gamma}^k\} = C(S, \gamma) \cap \text{spt} \|V_t\|$$

for all $t \in (-1, 1)$ *and*

$$\sup_{k+1 \le j \le n} \|f_j\|_{C^{1,\zeta}(B_{\gamma}^k \times (-1,1))} \le c(\mu + \|u\|_{L^{p,q}}). \tag{6.2}$$

Here $C^{1,\zeta}$ *is the norm in the parabolic sense.*

When $u = 0$, the assumption of Theorem 6.4 is the same as Theorem 6.1. In addition, when the norm $\|u\|_{L^{p,q}}$ is sufficiently small, the claim is that spt $\|V_t\|$ is a $C^{1,\zeta}$ graph near the center of the domain with the estimate (6.2). If $\nu = 0$ and V_t and u are independent of t, then just as in Sect. 5.1, we can prove that $0 = h + u^{\perp}$ for a.e. sense. We may regard $q = \infty$ in this case and (6.1) will be $p > k$, which is the same as Theorem 5.3. So we see a clear correspondence between Theorems 5.3 and 6.4. A remark is that the above theorem does not say anything extra about the meaning of "$v = h + u^{\perp}$". I do not know if the functions f_j have weak second derivatives in the space variables and weak first derivative in the time variable, for example. I suspect that they do exist and they satisfy $v = h + u^{\perp}$ for a suitable a.e. sense. If u is assumed to be in the Hölder class, then we have the following. The assumptions are repetitive but I write them out again to avoid any ambiguity.

Theorem 6.5 ([37, Theorem 3.6]) *Corresponding to* $E_1 \in [1, \infty)$, $\beta \in (0, 1)$ *and* $\nu \in (0, 1)$, *there exist* $\varepsilon \in (0, 1)$, $\gamma \in (0, 1)$, $c \in (1, \infty)$ *and* $T \in (2, \infty)$ *with the following properties. Suppose that the unit density* $\{V_t\}_{t \in (-T, T)}$ *in* $C(S, 1)$ *satisfies (i)–(iv) of Definition 2.2 and that it has a generalized normal velocity* $v(x, t) = h(V_t, x) + u(x, t)^{\perp}$. *Here* $u(x, t)^{\perp}$ *is the normal projection of* $u(x, t)$ *with respect to the approximate tangent space of* V_t *at* x. *Assume that*

(i) *for some* $t_1 \in [-T + 1/2, -T + 1]$, *we have* $\|V_{t_1}\|(\phi_S) \le (2 - \nu)\mathbf{c}$,

(ii) *for some* $t_2 \in [T - 1, T - 1/2]$, *we have* $\|V_{t_2}\|(\phi_S) \ge \nu\mathbf{c}$,

(iii) $\mu := \left(\int_{-T}^{T} \int_{C(S,1)} \mathrm{dist}\,(x, S)^2 \, \mathrm{d}\|V_t\| \mathrm{d}t \right)^{\frac{1}{2}} \le \varepsilon,$

(iv) *for all $B_r(x) \subset C(S, 1)$ and $t \in (-T, T)$, we have $\|V_t\|(B_r(x)) \le \omega_k r^k E_1$,*

(v) $\|u\|_{C^{\beta}} := \|u\|_{C^{\beta}(C(S,1) \times (-T,T))} \le \varepsilon.$

Then there exist $C^{2,\beta}$ functions $f_j : B_{\gamma}^k \times (-1, 1) \to \mathbb{R} \, (j = k + 1, \ldots, n)$ such that

$$\{(\hat{x}, f_{k+1}(\hat{x}, t), \ldots, f_n(\hat{x}, t)) : \hat{x} \in B_{\gamma}^k\} = C(S, \gamma) \cap \mathrm{spt}\,\|V_t\|$$

for all $t \in (-1, 1)$ and

$$\sup_{k+1 \le j \le n} \|f_j\|_{C^{2,\beta}(B_{\gamma}^k \times (-1,1))} \le c(\mu + \|u\|_{C^{\beta}}). \tag{6.3}$$

Here $C^{2,\beta}$ is the norm in the parabolic sense. Moreover, $v = h + u^{\perp}$ is satisfied pointwise on $C(S, \gamma) \cap \mathrm{spt}\,\|V_t\|$ for $t \in (-1, 1)$.

Because of the $C^{2,\beta}$ regularity of $\mathrm{spt}\,\|V_t\|$, we have the classical mean curvature vector and normal velocity defined at each point, and they satisfy the motion law in the classical sense. Once the regularity goes up this much, we have a well-defined parabolic PDE in the Schauder class. Thus, depending on the further regularity such as $C^{k,\beta}$ of u, $k \ge 1$, the standard parabolic regularity theory is applicable to have $u \in C^{k+2,\beta}$. In fact, even though I presented Theorem 6.1 first for the Brakke flow (the case of $u=0$), the proof of Theorem 6.1 depends on the establishment of Theorem 6.4 to get $C^{1,\zeta}$ and then Theorem 6.5 to get $C^{2,\beta}$ and then the parabolic regularity theory to go all the way to C^{∞}.

6.2   Outline of Proof for the Regularity Theorems

In this section, even though most of what is stated is not precise, some rough outline to prove Theorems 6.4 and 6.5 is presented. For simplicity, we consider the situation that $u = 0$, namely, we only consider the case of Brakke flow.

Just as in the Allard regularity theorem, a key is to obtain some "Caccioppoli inequality". In the case of stationary varifolds, we had that the first variation is equal to 0, and we used it to obtain (5.17), which is an analogue of (5.11). There, the analogue for $|\nabla f|^2$ was $\|T_x \Gamma - S\|^2$. For the Brakke flow, the inequality (2.11) does not give a direct access to $\|T_x \Gamma_t - S\|^2$ of moving Γ_t and we must think differently. In this sense, the regularity theory of Brakke flow requires a new perspective. It appears that this regularity theory utilizes more of the nonlinear structure than that of Allard. Just to get some idea, assume that $k = n - 1$ and $\mathrm{spt}\,\|V_t\|$ is represented as a smooth graph $x_n = f(\hat{x}, t)$ over a fixed $S \in \mathbf{G}(n, n - 1)$, where we identify

$\hat{x} = (x_1, \ldots, x_{n-1}) \in S$ and $\mathbb{R}^{n-1}$. Let ϕ_S be the function discussed above, which depends only on $\hat{x}$, and we use it in (2.11). The function ϕ_S does not have a compact support in $\mathbb{R}^n$ but we ignore this point for now. In this case, the left-hand side of (2.11) corresponds to $\frac{d}{dt} \int_S \phi_S^2 \sqrt{1 + |\nabla f|^2}$. Since we are concerned with the situation that f is close to S, we may roughly think that $|\nabla f| \approx 0$ and

$$\frac{d}{dt} \int_S \phi_S^2 \sqrt{1 + |\nabla f|^2} = \frac{d}{dt} \int_S \phi_S^2 (\sqrt{1 + |\nabla f|^2} - 1) \approx \frac{1}{2} \frac{d}{dt} \int_S \phi_S^2 |\nabla f|^2.$$

So we may pretend that the left-hand side is like the rate of change of Dirichlet energy of the graph. While this is so, it turns out that it is better not to linearize the left-hand side, and we will see the reason for this soon. The right-hand side of (2.11) may be written as

$$\int_S h \nabla \phi_S^2 \cdot \nabla f - \phi_S^2 h^2 \sqrt{1 + |\nabla f|^2} \approx \int_S 2\phi_S \Delta f \, \nabla \phi_S \cdot \nabla f - \phi_S^2 (\Delta f)^2,$$

where $h = \mathrm{div}\left(\frac{\nabla f}{\sqrt{1 + |\nabla f|^2}}\right)$ is the scalar mean curvature and we dropped the term of order $O(|\nabla f|^2)$, thus $h \approx \Delta f$. So the crude approximation of (2.11) is given by

$$\frac{d}{dt} \int_S \phi_S^2 \sqrt{1 + |\nabla f|^2} \leq \int_S 2\phi_S \Delta f \, \nabla \phi_S \cdot \nabla f - \phi_S^2 (\Delta f)^2.$$

We see that there is a $(\Delta f)^2$ term with negative sign, which is a "good term" since it generally has a dissipative effect on the left-hand side. By utilizing this term, we want to control the right-hand side by the L^2 norm of f, which is the quantity for which we have good control. By the Cauchy–Schwarz inequality, we continue as

$$\frac{d}{dt} \int_S \phi_S^2 \sqrt{1 + |\nabla f|^2} \leq \int_S -\frac{1}{2} \phi_S^2 (\Delta f)^2 + 2|\nabla \phi_S|^2 |\nabla f|^2. \tag{6.4}$$

By integration by parts, the last term may be estimated as

$$\int_S |\nabla \phi_S|^2 |\nabla f|^2 = -\int_S f \Delta f |\nabla \phi_S|^2 + 2f |\nabla \phi_S| \nabla f \cdot \nabla |\nabla \phi_S|$$

$$\leq \int_S \frac{1}{16} \phi_S^2 (\Delta f)^2 + 16 f^2 \frac{|\nabla \phi_S|^4}{\phi_S^2} + \frac{1}{2} |\nabla \phi_S|^2 |\nabla f|^2 + 2f^2 |\nabla |\nabla \phi_S||^2.$$

By moving the third term to the left-hand side, we obtain

$$\int_S |\nabla \phi_S|^2 |\nabla f|^2 \leq \int_S \frac{1}{8} \phi_S^2 (\Delta f)^2 + 32 f^2 \frac{|\nabla \phi_S|^4}{\phi_S^2} + 4f^2 |\nabla |\nabla \phi_S||^2. \tag{6.5}$$

Combining (6.4) and (6.5), we obtain

$$\frac{d}{dt}\int_S \phi_S^2\sqrt{1+|\nabla f|^2} \le \int_S -\frac{1}{4}\phi_S^2(\Delta f)^2 + 64f^2\frac{|\nabla\phi_S|^4}{\phi_S^2} + 8f^2|\nabla|\nabla\phi_S||^2. \qquad (6.6)$$

It turned out that we can obtain a very analogous inequality like (6.6) using (2.11). The reader can see the exact analogue of above computations in [21, p. 22 (5.54)–(5.58)], where T there corresponds to S in this book. The good news here is that the analogue of the last two terms of (6.6) turned out to be controllable using μ^2. More precisely, we can show an analogue of

$$\sup_{t\in[-T+1,T],\,x\in\text{spt}\,\phi_S} |f(x,t)|^2 \le c\int_{-T}^{T}\int_{S\cap B_1} f^2\,d\hat{x}dt (\approx c\mu^2),$$

and the last two terms of (6.6) can be bounded uniformly in time by $c\mu^2$. More precisely, see [21, Proposition 6.4], where one has

$$\text{spt}\,\|V_t\| \cap B_{4/5} \subset \{x : |S^{\perp}(x)| \le c\mu\} \qquad (6.7)$$

for all $t \in [-T+1, T]$, where c depends only on n, k. This formula is obtained via computations similar to Huisken's monotonicity formula and, as a side remark, it is essential that the "L^{∞} norm" is bounded linearly by μ when we obtain a decay estimate of μ. Continuing from above and with a serious abuse of notation, we obtain an analogue of

$$\frac{d}{dt}\int_S \phi_S^2\sqrt{1+|\nabla f|^2} \le c\mu^2 - \frac{1}{4}\int_S \phi_S^2(\Delta f)^2. \qquad (6.8)$$

Here is a crucial nonlinear estimate which is not precise or correct but is presented just to give a rough idea. If $\int_S \phi_S^2(\Delta f)^2$ is small and the area is close to a plane, we have an analogue of

$$\left|\int_S \phi_S^2(\sqrt{1+|\nabla f|^2}-1)\right| \le c\left(\int_S \phi_S^2(\Delta f)^2\right)^{\frac{n-1}{n-3}} + \text{lower–order term} \qquad (6.9)$$

for $n \ge 4$ and a better estimate for $n \le 3$. The precise statement corresponding to (6.9) is the following ([21, Proposition 5.2]).

Proposition 6.6 *Corresponding to $1 \le E_1 < \infty$ and $0 < v < 1$ there exist $0 < \alpha_1 < 1$, $0 < \mu_1 < 1$ and $1 < P_1 < \infty$ with the following property. For $S \in \mathbf{G}(n, k)$ and unit density varifold $V \in \mathbf{IV}_k(C(S, 1))$ with $\|V\|(C(S, 1)) < \infty$, define*

$$\alpha^2 := \int_{C(S,1)} |h(V,x)|^2\phi_S^2(x)\,d\|V\|(x),$$

$$\eta^2 := \int_{C(S,1)} |S^\perp(x)|^2 \, d\|V\|(x).$$

Suppose spt $\|V\|$ *is bounded and* $\|V\|(B_r(x)) \leq \omega_k r^k E_1$ *for all* $B_r(x) \subset C(S, 1)$.

(A) *Suppose*

$$\left| \|V\|(\phi_S^2) - \mathbf{c} \right| \leq \frac{\mathbf{c}}{8}, \quad \alpha \leq \alpha_1 \text{ and } \eta \leq \mu_1.$$

Then we have

$$\left| \|V\|(\phi_S^2) - \mathbf{c} \right| \leq \begin{cases} P_1(\alpha^{\frac{2k}{k-2}} + \alpha^{\frac{3}{2}}\eta^{\frac{1}{2}} + \eta^2) & \text{if } k \geq 3, \\ P_1(\alpha^{\frac{3}{2}}\eta^{\frac{1}{2}} + \eta^2) & \text{if } k \leq 2. \end{cases}$$

(B) *Suppose*

$$\frac{\mathbf{c}}{8} < \left| \|V\|(\phi_S^2) - \mathbf{c} \right| \leq (1 - \nu)\mathbf{c} \text{ and } \eta \leq \mu_1.$$

Then we have $\alpha \geq \alpha_1$.

This proposition itself has nothing to do with the Brakke flow and it is like a "Sobolev inequality", even though it is not. The idea is that the excess measure from a k-dimensional flat plane is controlled by the L^2 norm of the mean curvature raised by the power $\frac{2k}{k-2}$ plus lower–order terms when the measure is close to the flat plane and α and η are small. If the measure is away from a flat plane and η is small, then there is a positive lower bound for α. Just to check that this is believable, consider a sphere of radius r, even though it is not a graph. The $\mathcal{H}^k$ measure of such a sphere is $c(k)r^k$, while the L^2 norm of the mean curvature is $c(k)r^{\frac{k-2}{2}}$, so at least the exponent seems to match. If we consider a linearized quantity $\sqrt{1 + |\nabla f|^2} - 1 \approx |\nabla f|^2/2$, such an inequality does not hold, so the inequality seems genuinely nonlinear and geometric in nature. The proof of case (A) uses a Lipschitz graph approximation and somewhat lengthy. The line of proof of case (B) is simple as follows. Assume otherwise and suppose that we had a sequence $V_i \in \mathbf{IV}_k(C(S, 1))$ such that the corresponding α and η converge to 0 while the $\mathbf{c}/8 < |\|V_i\|(\phi_S^2) - \mathbf{c}| \leq (1 - \nu)\mathbf{c}$. Then by Theorem 1.15, we have a subsequential limit V which must be stationary and of the form $\theta|S|$ with an integer multiplicity θ. By Proposition 1.14, θ is constant and an integer. Since $\|V\|(\phi_S^2) - \mathbf{c}$ cannot be an integer, it is a contradiction.

For simplicity and just to get an idea, let us set the lower–order term of (6.9) to be 0. Write

$$E(t) := \int_S \phi_S(\hat{x})^2 (\sqrt{1 + |\nabla f(\hat{x}, t)|^2} - 1).$$

Then (6.8) and (6.9) give

$$\frac{\mathrm{d}}{\mathrm{d}t} E(t) \leq c\mu^2 - c E(t)^{\frac{n-3}{n-1}}$$

when the measure is close to the flat plane and $\alpha \leq \alpha_1$, and otherwise, we have

$$\frac{\mathrm{d}}{\mathrm{d}t} E(t) \leq c\mu^2 - c\alpha_1^2.$$

For the second possibility, we may assume $2c\mu^2 < c\alpha_1^2$, so $E(t)$ decreases at least for a fixed rate. Setting $\tilde{E}(t) := E(t) - c\mu^2 t$, we obtain

$$\frac{\mathrm{d}}{\mathrm{d}t} \tilde{E}(t) \leq -c\tilde{E}(t)^{\frac{n-3}{n-1}}. \tag{6.10}$$

Recapping what is stated above, if $\tilde{E}(-T + 1)$ is less than $(1 - \nu)\mathbf{c}$ but larger than $\mathbf{c}/8$, $\tilde{E}(t)$ will decrease at the rate of $c\alpha_1^2/2$ (or faster) until $\tilde{E}(t)$ becomes smaller than $\mathbf{c}/8$. Once there, $\tilde{E}(t)$ decreases either faster than $c\alpha_1^2/2$ or according to (6.10). Since the exponent $\frac{n-3}{n-1}$ is strictly less than 1, $\tilde{E}(t)$ will vanish in finite time. Thus $\tilde{E}(t)$ will be 0 for $t \geq -1$ as long as T is chosen appropriately large, and this means that $E(t) \leq c\mu^2$ is achieved after some elapse of fixed time. One can carry out the same kind of argument by reversing time. Since $E(t)$ is like the Dirichlet energy, we achieved a "Caccioppoli inequality". This is a very rough idea of getting the first estimate of the regularity theorem. In the actual proof, we obtain an estimate of $|\int \phi_S^2 \mathrm{d}\|V_t\| - \mathbf{c}|$ in terms of μ, that is, we obtain

$$\sup_{-1 \leq t \leq 1} \left| \int \phi_S^2 \mathrm{d}\|V_t\| - \mathbf{c} \right| \leq c\mu^2$$

(see [21, Theorem 5.7]). Since the change in time of $\int \phi_S^2 \mathrm{d}\|V_t\|$ bounds $\int \mathrm{d}t \int \phi_S^2 |h(V_t, x)|^2 \mathrm{d}\|V_t\|$ via (2.11), we obtain control of the L^2 norm of the Laplacian in terms of μ:

$$\int_{-1}^{1} \int \phi_S^2 |h(V_t, x)|^2 \mathrm{d}\|V_t\| \mathrm{d}t \leq c\mu^2. \tag{6.11}$$

At this point, we use (see [21, Lemma 11.2])

$$\int_{\mathbf{G}_k(U)} \|W - S\|^2 \phi_S^2 \, \mathrm{d}V(x, W)$$

$$\leq 4 \left(\int_U |h(V, x)|^2 \phi_S^2 \, \mathrm{d}\|V\| \int_U \mathrm{dist}(x, S)^2 \phi_S(x)^2 \, \mathrm{d}\|V\| \right)^{\frac{1}{2}}$$

$$+ 16 \int_U \mathrm{dist}(x, S)^2 |\nabla \phi_S(x)|^2 \, \mathrm{d}\|V\|,$$

which holds for any $V \in \mathbf{IV}_k(U)$. With this, (6.7) and (6.11), we have

$$\int_{-1}^{1} dt \int \phi_S^2 \|W - S\|^2 \, dV_t(x, W) \leq c\mu^2. \tag{6.12}$$

This is the tilt-excess estimate in terms of μ^2, which is the parabolic analogue of (5.16). The reader can see that it is a lot harder to obtain this than the stationary case.

With this estimate in hand, we may proceed with more or less the same idea as the case of Allard regularity theorem. We next show that ([21, Section 7]) there exists a parabolically Lipschitz function $\tilde{f} : B_{1/2}^k \times [-1/2, 1/2] \to \mathbb{R}^{n-k}$ with

$$\sup_{x_1, x_2 \in B_{1/2}^k, \, t_1, t_2 \in [-1/2, 1/2]} \frac{|\tilde{f}(x_1, t_1) - \tilde{f}(x_2, t_2)|}{|x_1 - x_2| + |t_1 - t_2|^{1/2}} \leq 1$$

and

$$\mathcal{H}^{k+1}((C(S, 1/2) \times [-1/2, 1/2]) \cap \cup_{t \in [-1/2, 1/2]} (\Gamma_t \triangle \, \text{graph} \, \tilde{f}(\cdot, t)) \times \{t\}) \leq c\mu^2. \tag{6.13}$$

Just like the proof for the Allard regularity theorem, (6.13) gives the error estimate of how the Lipschitz graph differs from Γ_t in terms of μ^2. This error estimate combined with (6.12) also gives

$$\int_{B_{1/2}^k \times [-1/2, 1/2]} |\nabla \tilde{f}(x, t)|^2 \, dx dt \leq c\mu^2, \tag{6.14}$$

and the L^∞ estimate (6.7) gives

$$\sup_{B_{1/2}^k \times [-1/2, 1/2]} |\tilde{f}(x, t)|^2 \leq c\mu^2. \tag{6.15}$$

The last step for $C^{1,\alpha}$ regularity is to obtain a decay estimate analogous to Proposition 5.7. For the precise statement, see [21, Proposition 8.1], where the reader can find a line-by-line analogy with Proposition 5.7. We proceed with a contradiction argument and assume that there exists a sequence of Brakke flows $V_t^{(i)}$ such that

$$\mu^{(i)} := \left(\int_{-T}^{T} \int_{C(W^{(i)}, 1)} |S^\perp(x)|^2 \, d\|V_t^{(i)}\| dt \right)^{\frac{1}{2}} \to 0$$

and $\|S - W^{(i)}\| \to 0$ as $i \to \infty$, with an appropriate set of other assumptions. Corresponding to each Brakke flow, we have a Lipschitz function $\tilde{f}^{(i)}$ satisfying (6.14) and (6.15). We then define

$$F^{(i)}(x, t) := \frac{\tilde{f}^{(i)}}{\mu^{(i)}}$$

and prove that the subsequential limit of $F^{(i)}$ is a solution of the heat equation. There are a few notable points in this argument different from the proof for the Allard theorem. Unlike Allard, where one can use the compactness theorem to obtain the strong L^2 convergence, we don't have any derivative control in time. On the other hand, there is a certain semi-decreasing property for the L^2 norm and this can be exploited to show the existence of strong L^2 convergent subsequence on $B_{1/2}^k \times [-1/2, 1/2]$. Also in proving that the limit function is the solution of the heat equation, we need to use non-negative test functions in (2.11). To do this, we use (6.15) to make sure that the substituted test function is non-negative. This aspect does not show up in the proof of Allard: in fact, all we need is (5.21) in terms of L^∞ estimate. Though all the details are omitted, these are the rough outlines of the proof of the $C^{1,\alpha}$ regularity theorem.

Even after we obtain the $C^{1,\alpha}$ regularity, this itself does not say anything about the velocity in the classical sense. This is also an interesting difference from the stationary case. There, once we have $C^{1,\alpha}$, the graph is a weak solution for a quasi-linear divergence form PDE with Hölder continuous coefficients, from which one can proceed with the linear PDE regularity theory. For the Brakke flow, we still have only the inequality (2.11) and I do not know any easy way out from this point. In [37], this problem of going from $C^{1,\alpha}$ to $C^{2,\alpha}$ is solved using the "second–order blow–up argument". The terminology is reflecting the fact that the $C^{1,\alpha}$ estimate uses an L^2 distance function from affine planes, while the $C^{2,\alpha}$ estimate uses an L^2 distance function from graphs of quadratic polynomial functions satisfying the heat equation. I only mention that there is a very special estimate [37, Proposition 4.3] which seems to be available only for the MCF since it essentially utilizes computations similar to those for Huisken's monotonicity formula. Hence, even if one obtains the $C^{1,\alpha}$ estimates as above, it is not a "standard linear parabolic theory" such as [23] which gives the $C^{2,\alpha}$ regularity.

6.3 Comments on the Regularity Results

It is obvious that the regularity statement for general Brakke flow cannot be better than stationary integral varifold. This is one reason that all the results in this chapter are stated under the assumption that the varifold is close to a unit density plane. When a stationary integral varifold V is close in B_1 to a union of three half-planes with unit density meeting along the boundary ("triple junction"), Simon [34] proved that the spt $\|V\|$ consists of three real-analytic surfaces with boundaries which meet along a real-analytic codimension–one surface near the origin. Simon's result is far more general than this but there are some technical conditions (which I do not go into) to apply his result for more general situations. In [38], we looked into the similar regularity result for Brakke flow in the case of $k = 1$. Namely, suppose that a one-dimensional Brakke flow V_t is close to a unit density triple junction in measure in some space–time neighborhood. Then spt $\|V_t\|$ consists of three $C^{1,\alpha}$ curves meeting at a point and moving regularly. The proof contains various new monotonicity-type estimates which are genuinely parabolic in nature. At present, there is a difficulty in generalizing the result to $k > 1$ but I hope that this can be solved in the near future. As I mentioned in Sect. 4.9, it is also interesting to investigate some regularity aspect of "dynamic stability" along with the existence results, even though it is not yet clear how to approach this aspect.

For Brakke flows which are obtained through the elliptic regularization method, White's local regularity theorem [42] gives the same type of regularity statement near the point of single multiplicity. The proof is much shorter than [21, 37], while having smoothness of approximating MCF is essential for the argument. In many geometric applications, construction through this path is convenient and it has been used widely. In comparison, one thing which [21, 37] lack is the "end-point" regularity statement typical of the parabolic regularity. Namely, we would like to say that the Brakke flow which is close to a flat plane in $B_1 \times (-T, T)$ is not only regular in the center of the space–time domain $B_\gamma \times (-1, 1)$, but in $B_\gamma \times (-1, T)$, assuming that $\|V_T\|(B_{1/2}) > 0$ for example. This aspect has been resolved by Lahiri [25] which proves exactly this. Lahiri also worked through the original proof of Brakke's regularity proof and provided a complete $C^{1,\alpha}$ proof [24] along the lines of Brakke.

References

1. Allard, W.: On the first variation of a varifold. Ann. Math. (2) **95**, 417–491 (1972)
2. Allard, W.K., Almgren, F.J., Jr.: The structure of stationary one dimensional varifolds with positive density. Invent. Math. **34**(2), 83–97 (1976)
3. Almgren, F.J., Jr.: Plateau's Problem: An Invitation to Varifold Geometry. W. A. Benjamin, Inc., New York-Amsterdam (1966)
4. Almgren, F.J., Jr.: Existence and Regularity Almost Everywhere of Solutions to Elliptic Variational Problems with Constraints. Memoirs of the American Mathematical Society, vol. 4, no.165. American Mathematical Society, Providence (1976)
5. Bellettini, G.: Lecture Notes on Mean Curvature Flow, Barriers and Singular Perturbations, Appunti. Scuola Normale Superiore di Pisa (Nuova Serie), vol. 12. Edizioni della Normale, Pisa (2013)
6. Bethuel, F., Orlandi, G., Smets, D.: Convergence of the parabolic Ginzburg-Landau equation to motion by mean curvature. Ann. Math. (2) **163**(1), 37–163 (2006)
7. Brakke, K.: The Motion of a Surface by Its Mean Curvature. Mathematical Notes, vol. 20. Princeton University Press, Princeton (1978)
8. Cheeger, J., Haslhofer, R., Naber, A.: Quantitative stratification and the regularity of mean curvature flow. Geom. Funct. Anal. **23**(3), 828–847 (2013)
9. Chen, Y.-G., Giga, Y., Goto, S.: Uniqueness and existence of viscosity solutions of generalized mean curvature flow equations. J. Differ. Geom. **33**(3), 749–786 (1991)
10. Colding, T.H., Minicozzi, W.P., II, Pedersen, E.: Mean curvature flow. Bull. Amer. Math. Soc. (N.S.) **52**(2), 297–333 (2015)
11. Ecker, K.: Regularity Theory for Mean Curvature Flow. Progress in Nonlinear Differential Equations and Their Applications, vol. 57. Birkhäuser Boston, Inc., Boston (2004)
12. Evans, L.C., Gariepy, R.: Measure Theory and Fine Properties of Functions. Textbooks in Mathematics, Revised edn. CRC Press, Boca Raton (2015)
13. Evans, L.C., Spruck, J.: Motion of level sets by mean curvature. I. J. Differ. Geom. **33**(3), 635–681 (1991)
14. Evans, L.C., Spruck, J.: Motion of level sets by mean curvature. IV. J. Geom. Anal. **5**(1), 77–114 (1995)
15. Giga, Y.: Surface Evolution Equations. A Level Set Approach. Monographs in Mathematics, vol. 99. Birkhäuser Verlag, Basel (2006)
16. Gilbarg, D., Trudinger, N.S.: Elliptic Partial Differential Equations of Second Order. Reprint of the 1998 edition. Classics in Mathematics. Springer, Berlin (2001)
17. Huisken, G.: Asymptotic behavior for singularities of the mean curvature flow. J. Differ. Geom. **31**(1), 285–299 (1990)

© The Author(s), under exclusive license to Springer Nature Singapore Pte Ltd. 2019 99
Y. Tonegawa, *Brakke's Mean Curvature Flow*, SpringerBriefs in Mathematics,
https://doi.org/10.1007/978-981-13-7075-5

18. Ilmanen, T.: Convergence of the Allen-Cahn equation to Brakke's motion by mean curvature. J. Differ. Geom. **38**(2), 417–461 (1993)
19. Ilmanen, T.: Elliptic regularization and partial regularity for motion by mean curvature. Mem. Amer. Math. Soc. **120**, 520 (1994)
20. Ilmanen, T.: Singularities of mean curvature flow of surfaces (1995). http://www.math.ethz. ch/~ilmanen/papers/sing.ps
21. Kasai, K., Tonegawa, Y.: A general regularity theory for weak mean curvature flow. Calc. Var. PDE. **50**, 1–68 (2014)
22. Kim, L., Tonegawa, Y.: On the mean curvature flow of grain boundaries. Ann. Inst. Fourier (Grenoble) **67**(1), 43–142 (2017)
23. Ladyženskaja, O.A., Solonnikov, V.A., Ural'ceva, N.N.: Linear and Quasi-linear Equations of Parabolic Type. American Mathematical Society (Translated from Russian) (1968)
24. Lahiri, A.: Regularity of the Brakke flow, FU Dissertationen Online (2014)
25. Lahiri, A.: A new version of Brakke's local regularity theorem. arXiv:1601.06710
26. Lin, F.H.: Some dynamical properties of Ginzburg-Landau vortices. Commun. Pure Appl. Math. **49**(4), 323–359 (1996)
27. Liu, C., Sato, N., Tonegawa, Y.: On the existence of mean curvature flow with transport term. Interfaces Free Bound. **12**(2), 251–277 (2010)
28. Mantegazza, C.: Lecture Notes on Mean Curvature Flow. Progress in Mathematics, vol. 290. Birkhäuser Springer Basel AG, Basel (2011)
29. Mantegazza, C., Novaga, M., Pluda, A., Schulze, F.: Evolution of networks with multiple junctions. arXiv:1611.08254
30. Metzger, J., Schulze, F.: No mass drop for mean curvature flow of mean convex hypersurfaces. Duke Math. J. **142**(2), 283–312 (2008)
31. Naber, A., Valtorta, D.: The singular structure and regularity of stationary and minimizing varifolds. arXiv:1505.03428
32. Schulze, F., White, B.: A local regularity theorem for mean curvature flow with triple edges. J. Reine Angew. Math. (to appear). arXiv:1605.06592
33. Simon, L.: Lectures on Geometric Measure Theory. In: Proceedings of Centre for Mathematical Analysis, Australian National University, vol. 3 (1983)
34. Simon, L.: Cylindrical tangent cones and the singular set of minimal submanifolds. J. Differ. Geom. **38**(3), 585–652 (1993)
35. Takasao, K., Tonegawa, Y.: Existence and regularity of mean curvature flow with transport term in higher dimensions. Math. Ann. **364**(3–4), 857–935 (2016)
36. Tonegawa, Y.: Integrality of varifolds in the singular limit of reaction-diffusion equations. Hiroshima Math. J. **33**(3), 323–341 (2003)
37. Tonegawa, Y.: A second derivative Hölder estimate for weak mean curvature flow. Adv. Calc. Var. **7**(1), 91–138 (2014)
38. Tonegawa, Y., Wickramasekera, N.: The blow up method for Brakke flows: networks near triple junctions. Arch. Ration. Mech. Anal. **221**(3), 1161–1222 (2016)
39. White, B.: Stratification of minimal surfaces, mean curvature flows, and harmonic maps. J. Reine Angew. Math. **488**, 1–35 (1997)
40. White, B.: The size of the singular set in mean curvature flow of mean-convex surfaces. J. Am. Math. Soc. **13**(3), 665–695 (2000)
41. White, B.: The nature of singularities in mean curvature flow of mean-convex surfaces. J. Am. Math. Soc. **16**(1), 123–138 (2003)
42. White, B.: A local regularity theorem for classical mean curvature flows. Ann. Math. (2) **161**(3), 1487–1519 (2005)
43. White, B.: Mean curvature flow (Stanford University), lecture note taken by Otis Chodosh. https://web.math.princeton.edu/~ochodosh/MCFnotes.pdf

Printed in the United States
By Bookmasters